Copyright © 2020 Joshua

All rights reserved.

No part of this book may be reproduced in any form or by any electronic or mechanical means, including information storage and retrieval system, without written permission from the author, excepted in the case of a reviewer, who may quote brief passages embodied in critical articles in a review.

Trademarked names appear throughout this book. Rather than use a trademark symbol with every occurrence of a trademarked name, names are used in an editorial fashion, with no intention of infringement of the respective owner's trademark. The information in this book is distributed on an "as is" basis, without warranty.

Although every precaution has been taken in the preparation of this work, neither the author nor the publisher shall have any liability to any person or entity with respect to any loss or damaged caused or alleged to be caused directly or indirectly by the information contained in this book.

The views expressed in this book are the author's own and do not necessarily represent those of the US Army, US Air Force, US Space Force, Department of Defense, or US Government.

Cover art courtesy of NASA archives.

ISBN-13: 9798655659230

Library of Congress Control Number: 2018675309
Printed in the United States of America

To my wife, Sha'Quana, and my daughters, Elliana, Annelise, and Madeline. I do this for them that they may live in a better world.

And to the men and women of the new United States Space Force, pathfinders to a new era of space development.

Contents

Foreword i
Acknowledgments vi
About the Author vii
Introduction viii
Chapter 1 **America's Silent Adversary** 1
Chapter 2 **Introduction to the Phases of Space Development** 13
Chapter 3 **Introduction to Astronautics and Spacepower** 27
Chapter 4 **Chinese Spacepower Theory** 46
Chapter 5 **American Spacepower Theory Today** .. 54
Chapter 6 **Chinese Space Strategy** 68
Chapter 7 **American Space Strategy** 86
Chapter 8 **The New Space Race** 93
Chapter 9 **Toward an American Space Strategy** 134
Chapter 10 **American Space Strategy Tomorrow** 143
Chapter 11 **The New Space Race, Reimagined** .. 156
Chapter 12 **Space Development Theory (SDT)** ... 174
Chapter 13 **Meeting the Future** 213
Chapter 14 **Conclusion** 231
Index 235
Table of Figures 242

Foreword

In 2020, America finds itself shaken by a pandemic, rattled by social unrest, wearied by long wars in the Middle East, and spiraling ever-deeper into national debt. As a subtext to all of this, we are waking up to evidence that behind the scenes we have endured continuous espionage, sabotage, subversions, and manipulations by hostile foreign powers. Those hostile powers are Communist China, and their loyal junior partner, Russia.

At the end of the First Cold War we welcomed China into the World Trade Organization believing that engagement would turn them quickly into a democratic partner. We failed to realize that we were welcoming a wolf in sheep's clothing into our midst and setting the stage for the Second Cold War. China has since drained trillions of dollars' worth of intellectual property from the West illegally, lured away our manufacturing sectors with unfair trade practices, embedded hundreds of thousands of their people as students in our universities and research centers, and engaged in ongoing misinformation campaigns designed to pit Americans against one another, as well as to disaffect our allies.

Worse still, an aggressive China is now seizing territory from its neighbors, using islands they themselves created in order to bolster their threadbare claims of ownership over various island chains. All the while they disregarded an international court's rejection of such claims. Adding to this disregard for international law, their earlier pledges not to weaponize the islands proved to be another diplomatic deception. Those islands now bristle with missiles and combat aircraft. Meanwhile, nearly two million Muslims endure the indignity of forced "reeducation" as captives in Chinese concentration camps.

Communist China has proven itself to be lawless, deceptive, and intolerant in their opportunistic quest to become the dominant world power by 2049, which marks the hundredth-anniversary of their Communist Revolution. Their behavior on Earth will be reflected in their behavior in space. Only America and its allies stand in its way.

While China's generals parade anti-satellite weapons in Tiananmen Square, Chinese and Russian diplomats at the United Nations parade a document titled the Prevention of the Placement of Weapons in Outer Space Treaty. Their charade of good will is nothing more than a brazen act of law-fare; an attempt to trick the West into agreeing to forego defending our space systems upon which our militaries, economic centers, and information driven societies depend. Acceding to such a treaty would open the West up to rapid defeat at the hands of a cheater. As already shown, China is a cheater.

Now squarely in the Second Cold War, author Joshua Carlson writes Spacepower Ascendant: Space Development Theory and New Space Strategy. This remarkable book does three things. First, it shares Carlson's vision of expanding the American economy into space to deliver the blessings of prosperity for generations to come. Second, it provides Carlson's Space Development Theory that offers a historically-based, step-by-step process for expanding America's economy beyond Earth's orbit to capitalize on what is expected to be a $10 trillion-dollar space economy by 2050. Finally, and at the core of this book, he compares and contrasts the American and Chinese space strategies, and then extrapolates how well they will perform against each other over the next thirty years. Spoiler alert: America, you must change your strategy or you will lose this Cold War to China and future Americans will be subordinated economically to decisions made in Beijing! Carlson provides his recommendations to adjust our strategy to prevail.

The vision, theory, and strategy laid out in this brief book should guide America's civil, military, and commercial space programs through the twenty-first century. Doing so will make us safer, more prosperous, and a better international partner. Perhaps more importantly, doing so will enshrine Western values and the rule of law as the guides for the peaceful development of outer space. Working together in a bipartisan spirit through the National Space Council, presidents must work with Congress to set in place an enduring legislative agenda that matches the vision contained herein.

There is a cautionary tale for the new Space Force. It must expand its roles and missions to protect, defend, and promote space commerce, while providing law enforcement to ensure all lawful and non-hostile users of space enjoy the same freedom of navigation that our Navy ensures on the high seas. Keeping a myopic focus on supporting terrestrial warfighters allows the Chinese the opportunity to seize the strategic points in space as well as the real estate with the highest commercial value. America must not let this happen.

This book will be of great interest to policy makers, space professionals, political scientists, and students of space development, spacepower, and strategic studies. It joins an expanding collection of spacepower theories that treat space as a new ocean, where the most important discoveries lie beyond Earth orbit, and have their greatest meaning in terms of economic development and human settlement. Success in the Second Cold War will be determined largely by our space policy and strategy.

I recommend this book be at or near the top of your reading list.

Stephen L. Kwast
Lieutenant General, USAF-Retired
San Antonio, Texas
20 June 2020

Acknowledgments

I wish to thank the many people that made invaluable contributions to this book. Foremost, I would like to thank Dr. Brent Ziarnick. He recommended I expand my original observations into book form. Without him, there would be no *Spacepower Ascendant*. I would like to thank Lt Gen (Ret.) Steve Kwast for kindly agreeing to provide the foreword to the book, it is an honor I do not take lightly. I would also like to thank Dr. Namrata Goswami, Dr. M.V. "Coyote" Smith, Dr. Peter Garretson, and Dr. Everett Dolman. They made their time and experience available to me, motivated by their love of space and their country. They expanded the book's scope to something worthy of publication. Additionally, I would like to thank Dr. Christopher Stone, Dr. Bradley Townsend, Maj Jose Negron, Maj Nic Schmidt, Mrs. Natasha Borg, and Ms. Chloe Marie Smith for their helpful and insightful comments during the review process. They helped take my attempts at grammar and make them intelligible.

About the Author

Joshua Carlson is a space acquisition professional and has served as an Air Force Government Civilian for the last 11 years, including nine years at the Space and Missile Systems Center, Los Angeles Air Force Base. At the same time, he is also a Captain in the California Army National Guard and deployed to a combat zone in the Middle East in support of Operation Inherent Resolve. A 2020 graduate of Air Command and Staff College, he concentrated his master's degree in space strategy as part of the Space Horizons Research Task Force. Currently, he lives in Texas with his wife and three daughters. Occasionally he finds time for his hobbies, which include archery, firearms, and theory development.

Introduction

This is a book that offers a unique look at two futures, two possible outcomes of the new space race. Either space is a place for the military alone, or it is a domain ripe with opportunities for industry as well.

How do you envision space?

Is it dark or full of wonder? Barren or lush with resources? A global common or somewhere that nations can continue the struggle of eons?

How we think about things and what they are limits what we can do with them. A stick, for instance, may be just a stick to one person. Another person may see an arrow, a drumstick, or a pencil. On the other hand, oxygen gas will never be an arrow, a drumstick, or a pencil no matter what you believe about it. Your vision determines what you see and what you do.

How do you envision space?

For decades, the US has seen space as a place of scientific *exploration* and military applications like reconnaissance, navigation aids, and communication. Due to technological and strategic limitations, that is all space has been. But that is not all it can be. Like the person locked into viewing a stick as a stick, or perhaps firewood, because they do not know it has other uses, the US is not conceptualizing space as it really is. It is dealing with space as it has been, not as it will be.

Not so China. China has envisioned space as the future lynchpin of their global empire and is taking steps to make that vision reality. Space has resources, energy, and area for massive industrial undertakings, if there is a vision to see it. China has that vision, and they are driving hard for it.

How do you envision space?

Vision is the key. Vision must drive technology and a strategy for the future. This book explores the visions of China and the US, and envisions a future based on those. This is no academic question, but something that will imminently impact the world regardless of national, cultural, or political background. The resources and capability that control of space offers will provide nearly unlimited mass to the nation or coalition that holds it. Solar power to run nations, iron to build another Earth, and water for settlement and propellant for a hundred generations. Right now, it looks like that coalition will either be headed by the US or China. Now is the time for action.

How do you envision space? Look up, the future of Earth is above.

The first chapter addresses the major competitor to the US, China, and some of their recent actions. The second chapter addresses the phases of space development, drawing a distinction between simply *exploring* somewhere and *expanding* national infrastructure to that location. The third chapter draws a delineation between astronautics and spacepower – between the civilian and industrial power in space, and the military applications.

The fourth and fifth chapters examine Chinese and US spacepower theories. Sixth and seventh examine the outworking of those theories, in the current Chinese and US space strategies. Chapter eight takes everything up to that point and puts it in a scenario to show what happens if the US pursues a more orbital-based strategy. Chapter nine reflects on the scenario and attempts to gain some lessons from it, while chapter ten proposes a new strategy. Chapter eleven re-runs the scenario with the new, proposed strategy and finds it successful in preventing Chinese victory.

Chapter twelve examines in-depth the proposed Space Development Theory to arrive at a detailed space theory for the US. Chapter thirteen looks to the future and makes some predications. Finally, chapter fourteen concludes with some parting thoughts on the new epoch that the world has entered.

The world has already changed, and the time to develop astronautics and spacepower is rapidly passing. Reaching for the stars too late will find them already occupied, the Second Cold War lost before it has barely begun.

Chapter 1

America's Silent Adversary

China's attitudes and aspirations towards expansionism, territoriality, and resource nationalism in space is of paramount significance to future space governance. China is a major spacefaring nation with specific future ambitions in space. Over the past several decades, China has witnessed rapid and observable progress with regard to its space activities.

--Namrata Goswami

China is the US' silent adversary and has been so for decades, even though it is only recently that they have felt able to directly challenge the US. China has taken the lessons from the US' rise in the late 19th century through maritime and seapower expansion. China envisions space in the same way.

The US emerged from WWII nearly unscathed and immediately pivoted from open combat with the Axis powers to a Cold War, resisting the spread of Communism. The struggle continued until the collapse of the Soviet Union in 1991. At that point, the US had emerged from nearly a century of conflict as the preeminent, singular, world superpower.[1] During the next decade, the US enjoyed unparalleled economic expansion and relative peace. The World Trade Center attack of 2001 changed the focus to terrorism, which absorbed the US' attention for nearly two decades. However, while the US's gaze was elsewhere, the world did not stand still.

China was continuing to grow, and forced the US to respond in 2011 with a 'Pivot to the Asia-Pacific,' aimed at containing China.[2] Despite that strategic shift, China has sustained its meteoric rise, surpassing the US purchasing power in 2013. It is currently the second-largest economy in the world by GDP and boasts a population roughly four times that of the US.[3] In 2007, in what was considered by many a shot across the US space bow, China tested an anti-satellite (ASAT) missile on one of its satellites, generating a debris cloud that still exists in orbit.[4] In a similar, but much less-publicized move, China landed a rover on the far side of the Moon in January 2019, the Chang'e-4 – a first in space exploration.[5]

While China's technology, resources, and possible intentions are concerning, its critical advantage rests in its strategy and innovation. Not technological innovation, which is often how the US uses the term, but ideational innovation – doing something new with something old, because of a new idea. The advantage that this innovation gives is time, because the something new being done does not have an automatic response, it takes longer to develop one. The US is beginning to realize just how hostile China is to US interests and the scope of their plans for space. A recent Congressional report specified, "China's goal...[is] to establish a leading position in the economic and military use of outer space... China views space as critical to its future security and economic interests...not merely to explore space, but to industrially dominate [it]."[6]

China has proclaimed that it will use space for only peaceful purposes.[7] It also proclaimed that it would use the islands it constructed in the South China Sea (SCS) for peaceful purposes before it chose to build military airbases and emplace electronic jamming equipment, anti-air and anti-ship missiles.[8] China has already proved its bad faith in dealing with the international community, and should not be trusted to suddenly become reliable in space.

As China pursues its goals in space, the irony is inescapable that the roots of China's space theory are American. Alfred Thayer Mahan wrote *The Influence of Sea Power Upon History* in 1890, in which he argued, from a historical perspective, that seapower would be decisive in the expansion of national power around the world.[9] In Mahan's definition, seapower goes beyond warships to include the personnel to crew them. Sustaining them is the industrial base and commercial benefits accrued.[10] At the time, the US was not the preeminent world power but rather a second tier behind the preeminent sea power, the United Kingdom. The First World War changed much of that calculus, bleeding the UK, and leaving Europe in ruins while the US escaped relatively unscathed. The Second World War completed the transformation, leaving the UK under siege for nearly two years, while the US became an economic powerhouse as the "Arsenal of Democracy." By the war's conclusion, the US had demolished the only Axis sea power, Japan, in devastating fashion. Since then, American fleets have served as the symbol of US power projection around the world, ensuring freedom of navigation and commerce. US naval dominance has been challenged in recent years by China, most dramatically by the seizure of the SCS in contradiction of the findings of the Permanent Court of Arbitration.[11]

The seizure of the SCS through the construction of fortified artificial islands is a strategy of *expansion* eerily reminiscent of the Chinese game of *Go*. For those unfamiliar with *Go*, the game begins with an open board, and players start placing pieces to secure terrain from the enemy. Unlike the tactically focused Chess, which only ends when one player captures the other's leader, *Go* allows a mix of strategies to be employed. These can range from aggressive moves to take terrain from the enemy to a more deterrence-focused occupation of territory anticipated to be critical. China's SCS strategy is to construct islands and put aircraft on them, essentially building unsinkable aircraft carriers to deter possible incursions from US naval assets, which looks very much like a deterrence *Go* strategy. As a result, it is unlikely that a US naval force would attempt to fight its way into the SCS during a military engagement.[12] In short, defeating the US strategy deters the US from acting. This countering is in keeping with the highest level of strategy, according to the Chinese masters. Sun Tzu recommends that the best victory is defeating the enemy's plan – without a bloody war.[13]

Both Mahan and Sun Tzu would recognize the Chinese strategy in the SCS, but it goes beyond that. China has already implied it will pursue a similar approach in space. In 2018, the head of the Chinese Lunar Exploration Program (LEP), Ye Peijian, said, "[t]he universe is an ocean, the moon is the Daioyu Island, Mars is Huangyan Island."[14] This statement is rich in meaning. First, he is stating that these two heavenly bodies are like islands that can be claimed and owned – and others *excluded*. This statement would be the same as the head of NASA proclaiming, "The moon is Hawaii and Mars is Guam." More disturbingly, the ownership of both islands is currently in dispute and they are not Chinese sovereign territory by international law. Daioyu, called Senkaku by the Japanese, has been the focus of heated exchanges between the two countries for decades. Likewise, Huangyan, called Pantang Shoal by the Philippines, was militarily seized by China in 2013.[15] China has subsequently declared a UN court ruling in favor of the Philippines void.[16] If space is viewed by China the same way that it sees those islands, then cooperation cannot occur since the Chinese government cannot be trusted to act in good faith. Competition is the only way forward, preserving superiority in the space domain to maintain economic dominance and national power.

The US is struggling to determine the way forward in that competition. Though the US Space Force (USSF) is now its own service, countering China in space cannot be done by a service focused only on support to Joint Warfighting here on Earth. A service must have a purpose larger than simply supporting another service. That complaint was one of the reasons why the US Army Air Corps demanded a different service run by 'air minded' leaders.[17] Now, the fledgling USSF has the mission of generating a purpose through control of the space domain and protection of space commercial interests. It remains to be seen whether the USSF will have the mental muscles to embrace this new mission. There is atrophy of those muscles and blame for that can be traced to the US Air Force (USAF). A recent article suggests the USAF has sought to conflate the air and space domains in order to attempt to retain the space mission for organizational benefit.[18] There is not even an approved definition of spacepower in Joint Publication 3-14, which seems odd for a publication dedicated to space operations.[19] As the USSF separates, it must strengthen its theoretical grasp on the domain and develop a mission of protection and furtherance of US industrial and commercial interests in space.

A common trope is that space is the "ultimate high ground."[20] While understandable and relatable to other services, this statement reveals terrestrial-centric thinking that the future USSF cannot afford and must leave behind like a spent booster stage. That kind of thinking has been useful, and likely necessary up to this point, but has exhausted its purpose.

It is short sighted to describe space as "high ground," just like it would be odd to refer to the ocean as "high ground." It focuses too much on Earth and ignores the strategic terrain of space. Both naval and space domains favor the projection of every element of national power and privilege maneuver. To set the proper tone for the new USSF, and to break its mindset away from the tyranny of this terrestrial perspective, a better moniker for space would be "The Silent Sea." It encompasses the maritime and naval analogies of space *expansion*, as well setting the space domain apart as unique.

As America enters the third decade of the twenty-first century, China is the primary competitor, and the decisive area of conflict will be space. Space services will reshape the global economy in the next two decades, surging to between $1T and $2.7T in economic output by 2040.[21] While China has outlined its strategy for at least the next thirty years, there is no similar outline from American political or space leaders. This book provides a recommended vision for the USSF and Space Operations Command that can drive their fledgling mission statements and goals as they evolve. That vision will then be projected into the future, out to the dawn of the next century, to explore spacepower capabilities and opportunities. Because the subtext of this book is competition with China, their own goals and vision will be projected alongside for analysis of differences and outcomes. Before examining the extrapolation of both strategies, this book will discuss Space Development Theory (SDT) and Astronautics vs. Spacepower to provide a base for discussion. It is time to start the journey.

---

[1] Tom Engelhardt, "The US Has Been the World's Sold Superpower for the Last 13 Years – Why Hasn't It Done Anything Good?" Published 16 September 2014. Accessed 15 March 2020.
https://www.thenation.com/article/archive/us-has-been-worlds-sole-superpower-last-13-years-why-hasnt-it-done-anything-good/

[2] Mark E. Manyin, et al, "Pivot to the Pacific? The Obama Administration's 'Rebalancing' Toward Asia" Congressional Research Service, 28 March 2012. Accessed 21 May 2020: https://fas.org/sgp/crs/natsec/R42448.pdf

[3] Knoema, "The World's Largest Economy: China vs United States" https://knoema.com/fsvntfc/the-world-s-largest-economy-china-vs-united-states (Accessed 15 March 2020).

[4] Leonard David, "China's Anti-Satellite Test: Worrisome Debris Cloud Circles Earth" *Space.com*. 2 February 2007. https://www.space.com/3415-china-anti-satellite-test-worrisome-debris-cloud-circles-earth.html. Accessed 15 March 2020.

[5] Sarah Pruitt, "China Makes Historic Landing on 'Dark Side' of the Moon" History.com. 3 Jan 2020. Accessed 20 February 2020. https://www.history.com/news/china-plans-historic-landing-on-dark-side-of-the-moon.

[6] Congress, "2019 Annual Report," US-China Economic and Security Review Commission, Washington D.C, November 2019, 359.

[7] State Council Information Office, "White Paper on China's Space Activities", People's Republic of China, 2000, 2006, and 2011. *http://www.scio.gov.cn/zfbps,*

[8] "Chinese Land Reclamation in the South China Sea: Implications and Policy Options," Congressional Research Service, Washington D.C., 18 June 2015. Accessed 15 June 2020: https://www.everycrsreport.com/files/20150618_R44072_f366ec875f807562038948748386312c12acd5f4.pdf; Kyle Mizokami, "China's South China Sea Military Bases Are More Than They Seem" *The National Interest*. 11 January 2020. https://nationalinterest.org/blog/buzz/chinas-south-china-sea-military-bases-are-more-they-seem-112706. Accessed 15 March 2020.

[9] A. T. Mahan, *The Influence of Sea Power upon History, 1660-1783* (New York: Dover Publications, 1987), 26–29.

[10] Mahan, 28.

[11] "PCA Press Release: The South China Sea Arbitration (The Republic of the Philippines v. the People's Republic of China," Permanent Court of Arbitration, The Hague, the

Netherlands, 12 July 2016. Accessed 15 June 2020: https://pca-cpa.org/en/news/pca-press-release-the-south-china-sea-arbitration-the-republic-of-the-philippines-v-the-peoples-republic-of-china/.

[12] Kathy Gilsinan, "How the U.S. Could Lose a War With China" *The Atlantic*. 25 July 2019. https://www.theatlantic.com/politics/archive/2019/07/china-us-war/594793/. Accessed 15 March 2020.

[13] Sun Tzu, *The Art of War*, trans. Samuel B. Griffith (Oxford: Oxford University Press, 1971), 77.

[14] Brendon Hong, "China's Looming Land Grab in Outer Space", *Daily Beast,* June 22, 2018 at https://www.thedailybeast.com/chinas-looming-land-grab-in-outer-space (Accessed 22 January, 2020).

[15] Scott Neuman, "Little Islands Are Big Trouble in the South China Sea", *NPR*, 7 September 2012 at https://www.npr.org/2012/09/07/160745930/little-islands-are-big-trouble-in-the-south-china-sea. (Accessed 15 March 2020).

[16] Tom Phillips, Oliver Holmes, Owen Bowcott, "Beijing rejects tribunal's ruling in South China Sea case", 12 July 2016 at https://www.theguardian.com/world/2016/jul/12/philippines-wins-south-china-sea-case-against-china. Accessed 15 March 2020.

[17] Maj Chris Wachter, "Air-Mindedness: The Core of Successful Air Enterprise Development," Air & Space Power Journal, January-February 2012, 1.

[18] Brad Townsend, "Space Power and the Foundations of an Independent Space Force", Air & Space Power Journal, Winter 2019. 12-13.

[19] Townsend, "Space Power", 12.

[20] Benjamin S. Lambeth, *Mastering the Ultimate High Ground: Next Steps in the Military Uses of Space*. (Santa Monica, CA: RAND Corporation, 2003.) https://www.rand.org/pubs/monograph_reports/MR1649.html; Colin Drury, "US space force: Trump says 'we will control the ultimate high ground' as he launches new military service" Independent.uk. Published 21 December 2019. Accessed 20 February 2020.

https://www.independent.co.uk/news/world/americas/us-space-force-launch-trump-military-budget-2020-us-air-force-a9256051.html

[21] The Future Space Economy, Morgan Stanley. 2 July 2019 at https://www.morganstanley.com/ideas/investing-in-space. Accessed 15 March 2020; The space industry will be worth nearly $3 trillion in 30 years, Bank of America predicts. CNBC. https://www.cnbc.com/2017/10/31/the-space-industry-will-be-worth-nearly-3-trillion-in-30-years-bank-of-america-predicts.html

For more reading on China and its strategy, see:
- Howard French, *Everything Under the Heavens: How the Past Helps Shape China's Push for Global Power,* Knopf, Borzoi Books, New York, New York, 2017.
- Christopher Stone, *Reversing the Tao: A Framework for Credible Space Deterrence,* CreateSpace Independent Publishing Platform, 9 June 2016
- Peter Navarro, *Death by China: Confronting the Dragon – A Global Call to Action,* Pearson FT Press, Upper Saddle River, New Jersey, 2011.
- Robert Spalding, *Stealth War: How China Took Over While America's Elite Slept,* Portfolio/Penguin Publishing, 1 October 2019.
- Jim Sciutto, *The Shadow War: Inside Russia's and China's Secret Operations to Defeat America,* HarperCollins Publishers, 14 May 2019.
- Newt Gingrich, *Trump vs. China: Facing America's Greatest Threat,* Center Street Publishing, 22 October 2019.
- Michael Pillsbury, *The Hundred-Year Marathon: China's Secret Strategy to Replace America as the Global Superpower,* St Martin Publishing, New York, NY, 2016.

- Jonathan Ward, *China's Vision of Victory*, Atlas Publishing and Media Company, LLC, Printed in the United States, 2019.
- Oriana Skylar Mastro, "The Stealth Superpower: How China Hid Its Global Ambitions," Foreign Affairs, January/February 2019. Accessed 11 June 2020: https://www.foreignaffairs.com/articles/china/china-plan-rule-asia.

Chapter 2
Introduction to the Phases of Space Development

This time, we will do more than plant our flag and leave our footprints, we will establish a long-term presence, expand our economy and build the foundation for the eventual mission to Mars, which is actually going to happen very quickly.

--President Donald J. Trump

An introduction to the phasing of Space Development Theory, it offers a vision of what space development should look like and how it would progress. This serves to undergird both scenarios discussed later in the book and illuminates why China's current strategy is so dangerous. Domain development progresses through the steps of *exploration, expansion, and exploitation,* with occasionally required *exclusion* to prevent hostile encroachment.

People go places and do things for reasons. Space Development Theory (SDT) attempts to systematize these places, things, and purposes into a structure that categorizes past, present, and future actions. SDT breaks development of space into four phases: *exploration, expansion, exploitation,* and *exclusion. Exploration* is going to a new domain and seeking information about it that could be strategically important, such as key locations or resource nodes. *Expansion* is extending national architecture to control key locations and resources, providing a permanent presence. *Exploitation* is when the nation experiences a net benefit from the new domain and resources, growing national power. Finally, *Exclusion* is national ownership and protection of an area from hostile attempts at *exploration, expansion,* or *exploitation.* These will be expanded upon later in this chapter.

All missions are not equal, and all purposes are not the same. In a recent speech by the head of NASA, Jim Bridenstine responded to China's rover landing on the far side of the moon by contrasting NASA's landing on Mars. He responded to the criticism that the US had fallen behind technologically with, "We did not fall behind...we landed on the far side as well...*of Mars.*"[1]

While this is a useful rhetorical response, it falls short on further examination. The problem with this statement is that the comparison is false – a mission is not *just* a mission – the purpose and intention of that mission are important. The US mission to Mars is "doing science," as he notes later, but China's mission is much more than that.[2] China's mission does science as well, bringing seeds from two plants along to sprout them in microgravity.[3] However, the intent for that science is very different. Liu Hanlong, chief director of the experiment, was explicit on his purpose, "our experiment might help accumulate knowledge for *building a lunar base and long-term residence on the moon.*"[4] A mission to the Moon for *expansion* is not equivalent to a mission to Mars for *exploration.* While Administrator Bridenstine can undoubtedly be forgiven for his attempt to use rhetoric to downplay China's accomplishment, it does not change the fact that his comment in this case unhelpfully generalizes space missions.[5] The United States has led the world's *exploration* of space for more than forty years. However, it has never *expanded* beyond Earth's geostationary orbit. The danger of China's strategy is that they intend to *expand* to the Moon as early as 2036, and the US struggles to have the mental framework to understand this strategy.[6]

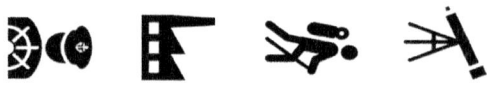

- Exploration (E1)
  - Entry – physical proximity to the location, able to observe, map, and evaluate
  - Actions – identify key locations and resources and phenomena impacting access and exploitation
  - Exit – physical presence and permanent facilities, claims of national interest
- Expansion (E2)
  - Entry – able to sustain presence, either permanent or rotational
  - Actions – establish logistics, observation, and communication outposts, forts, and living areas; either military or economic leading expansion
  - Exit – area in question and logistic lane are secure enough to begin exploitation; net economic benefit to the nation is positive
- Exploitation (E3)
  - Entry – projected gain is high enough, and risk to commercial interests low enough, for commercial interests to
  - Actions – establish gathering, production and distribution nodes
  - Exit – resource can be exhausted or hostile forces could project sufficient force to cause supposition; if promotion, nation considers area key national interest
- Exclusion (E4)
  - Entry – conditional phase, requires contested control by hostile powers and area viewed ideationally as a member of the national body; may be driven through either political or popular factors
  - Actions – may be offensive or defensive; aims to eliminate hostile occupation or prevent hostile occupation of friendly terrain through denial of occupation, sustainment, or transit; extensive cooperation between military and economic instruments, with military being supported by economic
  - Exit – the contested control is eliminated, either through victory or defeat and focus will shift to a different phase until control is contested again

Figure 1 - Phases of Space Development (original by author)

SDT posits the following phases of development, described by the "Four E's": *exploration, expansion, exploitation,* and *exclusion.* Figure 1 above provides a summary of the phases and entry/exit criteria, as well as some actions done during the phases. These form a framework consistent with other domains, and the common thread of humanity ties them together. While the first three are genuinely serial, *exclusion* can occur at any point. It may even repeat several times during the other phases. *Exclusion* is ideational and human-driven, while the others are all a striving of humanity against the environment.

Each phase has roles for the instruments of national power, which are divided into diplomatic, informational, military, and economic categories. Some definitions will prove useful in understanding the roles they play in space development. The diplomatic tool is defined as, "the principal instrument for engaging with other states and foreign groups to advance...[national] values, interests, and objectives, and to solicit foreign support for...[national] military operations."[7] The informational tool is defined as, "using information to further...[national] causes and undermine...[hostile ones]."[8] The military tool will be defined differently than US joint doctrine, and will be expanded to, "all military organizations or capabilities that can control a domain while threatening to destroy or degrade competing forces or factions." Finally, the economic tool is defined as, "the sum of all economic output, as well as trade and industrial capabilities."

The first step in SDT is conducting *exploration* to understand the new domain and specific operating areas. *Exploration* includes identifying resources and key military terrain, understanding the domain elements that make it unique, and recording and systematizing these findings for use by individuals and national organizations. *Exploration* provides significant information to a nation, and so the informational elements of national power benefit considerably from it. However, depending on the location to be explored, diplomatic treaties may need to be negotiated before *exploration* can begin. The military instrument of power will likely need to be involved in some way, as militaries train to move in austere or hostile environments. Finally, areas of potential economic benefit are identified in this phase. However, actual benefits are minimal and likely limited to providing unique equipment designed for *exploration*. Historical examples of this would be the Lewis and Clark expedition, the US visits to the Moon, and the many naval circumnavigations of the globe.[9]

Second, *expansion* is conducted by the movement of some combination of economic and military forces into the area to establish enduring access to those previously identified key terrain and resource nodes. Focus of *expansion* is securing access through occupation, and not *excluding* competitors from the area entirely. *Expansion* includes patrols, settlements, the building of bases, both forts and servicing stations, and possibly altering the terrain to make it more advantageous. *Expansion* requires the projection of enduring forces into the region. Whatever is present must have some capability of sustainment and self-defense to avoid destruction from the environment or hostile powers, especially if the position or resources at stake are of strategic benefit. *Expansion* cannot be accomplished merely through denial from another adjacent domain.[10] For *expansion* to be successful, national power and legal framework must be enforceable in the new location.[11] Historical examples of this would the US Cavalry expanding American influence into the Midwest, the race to settle the Americas, and the British occupation of Singapore and Gibraltar.

Third, *exploitation* moves the focus from holding an area and enforcing ownership to bolstering the economic power of a nation. The *exploitation* phase begins when the area becomes economically net positive to the country in question. That is, the cost of infrastructure and holding forces are offset by economic benefits gained. *Exploitation* includes gathering of needed resources, the establishment of trade lanes, the development of new markets for discovered resources, and manufacturing at identified sites in the new area. National economic power benefits most from economic growth in this phase and especially control of strategic resources.

Mahan cites the British for their superior *exploitation* over Spanish and Portuguese attempts.[12] In contrast to the British, who built self-sustaining settlements, the Spanish and Portuguese simply stripped resources and then returned to their home country. Britain grew its *expansion* and *exploitation* in synergy while the other two did not. This resulted in their New World claims being bought by their own gold, which had made its way to Britain through superior economic positioning.[13]

Of note, this phase can be considered the most important as the path to hegemony begins with dominating the production of the most valuable commodities.[14] Indeed, it is possible to become the new hegemon without the need for a costly and risky direct conflict. If one nation dominates the production and shipping of valuable resources, and that nation then focuses on becoming the lender to the world, it will likely become a hegemon.[15] This strategy can be seen as perfect indirect warfare, becoming the leader without needing to fight for it. War is still possible, and conflict likely, as no nation will willingly surrender hegemony. However, the country seeking hegemonic status does not necessarily need a war to accomplish it.

Returning to the phases of development, fourth is *exclusion*. As the region becomes more central to a country's plans and strategic outlook, the nation will naturally seek to *exclude* other powers from being able to interfere in it. A nation will seek exclusive access, ultimately seeking to incorporate it into the nation's political 'body' if it is important enough. *Exclusion* usually requires the full spectrum of national instruments of power. The diplomatic instrument negotiates for international recognition. Information emphasizes the importance of the region to the nation, seeking to convince both internal and external audiences. The military conducts patrols and basing. The economic instrument conducts embargoes to impact the economy of other nations antagonistic to claims of sovereignty. However, *exclusion* does not have to be active to be effective. Passive *exclusion*, like the threat weaponized islands in the SCS present to the US Navy, is more effective than direct conflict because a rational actor sees the cost of interference as too high and will not attempt it. Historical examples for both active and passive *exclusion* include the US *exclusion* of foreign powers from the Americas via the Monroe Doctrine and the British exclusion of German shipping from the North Sea during both World Wars.

The phases of *exploration, expansion, exploitation,* and *exclusion* comprise the structure of space development and provide insight into how the new space race will develop. Nations seeking to expand into any new domain have followed this same process, and space will be no different. These phases are severable by location, are not linear, are not inherently hostile, nor are they absolute.

Severable locations mean that different areas may be in different phases. A nation may be *exploiting* the cislunar area while *exploring* Mars. Each successive area must undergo these phases, and clarity on what to define as 'areas' must even occur before this can begin. For instance, solar power is only a viable energy source out to the asteroid belt. Designating everything between the Sun and asteroid belt as inner space and *exploring* that area separately from outer space would make sense. Further *exploration* will help define what 'inner' and 'outer' are, where the line is, and if it shifts over time. Similarly, *exploration* has discovered a lot about Earth's orbits and magnetosphere, but significantly less about the other planets. *Exploration* of Earth is not likely required, but Uranus would need to be *explored* in some detail to bring it up to similar levels of familiarity.

Non-linearity means that a society may need to *explore,* and a company inside the nation may race to *exploit* a particularly valuable resource before that nation has *expanded* into that region. Alternatively, a country that has been *exploiting* an area could lose it due to war and, a decade later, after it has recovered it, may need to *explore* it again to determine if a significant change has occurred. While they are not always linear, they generally follow the outlined progression. If the discovery of a new domain allows only one nation or group access to it, they would still need to *explore, expand, and exploit* in that order. It would likely not need to *exclude*, as that would require a human threat to its ownership. This scenario is theoretical, however, and no group encountering a new domain can know that it is uncontested. Therefore, they must *explore,* and the process begins as laid out.

The phases not being inherently hostile means that friendly nations or alliances could be allowed, or even invited, to participate in any of the phases. Also, several friendly nations could participate as equal partners in an *expansion* or *exploitation,* so their pooled resources benefitted all nations involved. Indeed, this sharing of risks and benefits may serve as a critical building block of the alliance and serve to benefit all parties. The *exclusion* is not of all non-national parties. Britain in WWII, as was already noted, *excluded* German shipping through Gibraltar while allowing passage to non-hostile. Likewise, the current US dominance in naval assets in the world's oceans has allowed greater freedom of maneuver for most nations and served as a basis for international cooperation in pursuit of mutually beneficial goals.[16] *Exclusion*, the only phase oriented on the human terrain, is inherently competitive but variably hostile – it can range from embargos to outright armed conflict. *Exclusion* is conducted against rival powers and not allied or neutral actors.

The phases are not absolute, *exploration* and *expansion* can occur concurrently. Alternatively, the *exploitation* phase may have been reached and new sensor development revolutionizes the analysis of space, and that drives a further *exploration*. The question is of national drive, energy (or national capability and resources), and competition. High national drive and energy, with low competition, could allow a nation to engage in several of the phases at once. High competition would force the nation that has already secured the area to slow down and spend more time carefully *excluding* to avoid losing the investments to infrastructure made during *expansion*. Alternatively, if the nation has not *expanded* to the key areas and does not have sufficient strength to *exclude* hostile powers, then it may move faster to ensure its positional dominance. Also, if either national drive (or will) is low, then even high energy and low competition may not encourage aggressive multi-phase development. Finally, high national drive and low competition, if not paired with sufficient national energy, will simply not allow sufficient resources to engage in multiple phases simultaneously.

The US has been in space since the 1950s. However, most work and development has not progressed beyond the *exploration* phase. The US seems to think that it is somehow in a superior position and, while the capability present in orbit is impressive, there is no *expansion* beyond Earth orbit and virtually nothing *exploited.* The US must seek to *expand* and *exploit* space with all speed because domains do not remain unclaimed for long. Humanity will develop space, and if the US has not engaged in the required *expansion* and *exploitation*, then the US' proud history in space will be just that. History, confined to books.

---

[1] NASA Video, *Administrator Bridenstine speaks with employees at NASA's Glenn Research Center,*
(Youtube.com, video) 10 July 2019. (timestamp 2:40-3:33).

[2] NASA Video, *Administrator Bridenstine.*

[3] Rafi Letzter, "There are Plants and Animals on the Moon Now (because of China)", LiveScience, January 3, 2019. Accessed 25 May 2020:
https://www.livescience.com/64413-china-space-moon-plants-animals.html.

[4] Letzter, "Plants and Animals on the Moon" (emphasis added)

[5] For background on Bridenstine's thoughts on space and expansion, see: Jim Bridenstine, "This is out Sputnik Moment," OKGrassroots.com, 6 November 2016. Accessed 11 June 2020: https://okgrassroots.com/?p=642815.

[6] Lifang, "China Focus: Flowers on the Moon? China's Chang'e 4 to Launch Lunar Spring", *Xinhuanet*, April 12, 2018. Accessed 25 May 2020:
http://www.xinhuanet.com/english/2018-04/12/c_137106440.htm

[7] Joint Chiefs of Staff, "Joint Publication 1: Doctrine for the Armed Forces of the United States," March 25, 2013, I-12,
https://www.jcs.mil/Portals/36/Documents/Doctrine/pubs/jp1_ch1.pdf.

[8] Joint Chiefs of Staff, "JP 1.", I-12.

[9] History.com Editors, "Lewis and Clark Expedition," History.com, 31 January 2020. Accessed 16 June 2020:

https://www.history.com/topics/westward-expansion/lewis-and-clark.
[10] Peter Garretson, "USAF Strategic Development of a Domain," Over The Horizon Journal, 10 July 2017. www.othjournal.com/2017/07/10/strategic-domain-development/, 3.
[11] For a discussion of some legal issues intrinsic in space *expansion*, see Ceren Goncu, "Reflections Of Deep Space Activities on International Relations," Izmir University of Economics, Turkey, July 2019. Accessed 11 June 2020: https://www.academia.edu/40679299/Reflections_of_Deep_Space_Activities_on_International_Relations.
[12] Mahan, *The Influence of Sea Power upon History, 1660-1783*, 56.
[13] Mahan, 56.
[14] Garretson, "USAF Strategic Development", 14.
[15] Garretson, 14.
[16] National Security Council, "Countering Piracy off the Horn of Africa: Partnership & Action Plan" United States of America, December 2008, 7.

Chapter 3
Introduction to Astronautics and Spacepower

I don't think the human race will survive the next thousand years, unless we spread into space. There are too many accidents that can befall life on a single planet. But I'm an optimist. We will reach out to the stars.

--Stephen Hawking

Up to this point, most of the things done in space have simply been called 'spacepower' – but that does not consider the difference between military and civilian applications. Space has been able to escape making this separation until now because civilian presence was virtually non-existent. That is no longer the case and will become glaringly obvious in the next two decades. To account for this, the term astronautics is introduced to describe the civilian and industrial presence in space.

There have been many previous writers that have described their concepts of spacepower theory. Brent Ziarnick described spacepower in terms of two deltas – grammar and logic – that build and project spacepower into the domain.[1] Everett Dolman described spacepower in terms of *astropolitik*, taking and holding the key areas of space. Neither of these provide sufficient delineation between civilian presence and military presence in space. Further development of domain understanding would be useful. Specifically, breaking the elements of power into a military component and an all-encompassing component, like more mature domains.

While most authors group all power in space under the heading of *spacepower,* it is helpful to separate *spacepower* from *astronautics.* This delineation follows the trend of the non-land domains. The Navy recognizes the difference between seapower and maritime The Air Force similarly distinguishes airpower from aviation. *Spacepower* is defined as military force that can exert influence in and from the domain and create effects in other domains for strategic benefit. *Astronautics* is defined as those elements that are primarily commercial and industrial; it includes all aspects that allow for projection into, production, sustainment, training, profit, and expansion in the domain for the purpose of strategic benefit. These would be related to the terms of military power and latent power in the realist theory of international relations, respectively. Spacepower is military power in the space domain. Astronautics is industrial and commercial, adding to or enhancing a nation's potential in the domain and provides for the projection of military forces. More mature domains recognize the importance of this division, and space must use a similar construct. Figure 2 below shows a conceptualization of astronautics and spacepower.

Astronautics
Those elements that are majority commercial or industrial in nature and include all elements that allow for projection into, production, sustainment, training, profit, and expansion in the domain.
(See also Maritime or Aviation Power)

Spacepower
Military force that can exert influence in and from the domain or create effects in other domains. (See also Seapower or Airpower)

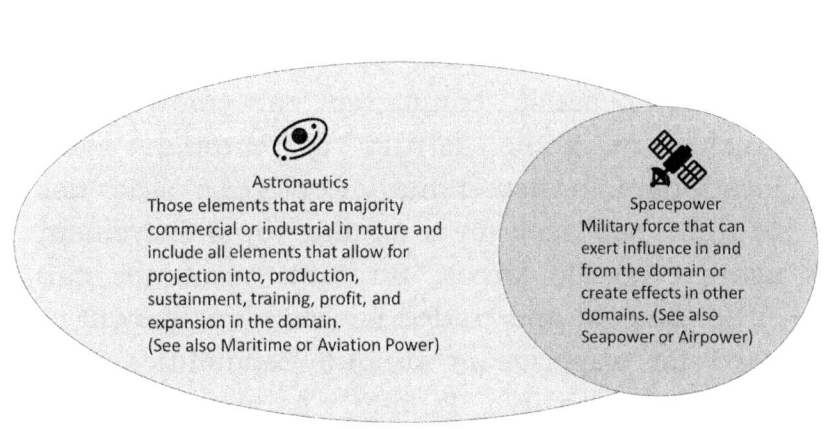

Figure 2 - Astronautics and Spacepower (original by author)

Fundamental to both definitions is the concept of strategic benefit – which is anything found in a domain that increases economic or military national capability and assists in achieving a nation's strategy in a significant fashion. The most important part of the definition is related to the grand strategy itself, which changes over time and encompass enemy and friendly stratagems. Friendly strategies should focus on securing strategic benefits for themselves and their allies, while generally denying them to hostile powers. The other element of the definition is that it assists economic or military capability. These can be strategic locations, such as the Straits of Gibraltar, strategic resources, such as oil, fish, or rare earth metals, or strategic knowledge, such as knowledge of trade winds or currents. Locations, resources, and knowledge can all have both economic and military applications, depending on the resources and the strategies of the nations involved. Locations can be both economic and military because commerce can move more quickly in some terrain or through some locations. Likewise, military forces can either use or deny passages for a similar rapid movement against hostile forces. Strategic resources can either be sold and traded for economic benefit or used for superior or unique capabilities in a military capacity. Finally, gaining strategic knowledge requires human interaction with the domain. Commercial uses would be the rapid movement of ships on the currents that enabled quicker voyages, and more money to the owning nation. Military units could use the same currents to outpace a hostile force.

Astronautics covers technology, industry, personnel, commerce, and basing. While the technology appears obvious, it is nonetheless valuable to highlight in both its absolute and relative senses. In an absolute sense, without the requisite technology, space travel and space *exploitation* are impossible. Astronautics got its start with Dr. Robert Goddard, considered the father of modern rocket propulsion. He experimented with rockets in 1926 and was the first to fly a rocket using liquid propellant.[2] Of course, technology is ambivalent to the purpose of its task; the same technology that was designed to bring people to the Moon instead brought death to the people of London in the incarnation of V-2s. This highlights the truth that all technology can cause destruction through human application. Even technologies with the best of intentions can become terrible weapons in the wrong hands.

Balancing this pessimistic truth is that, because technology is a tool, it can also accomplish much good. Dr. Wernher von Braun, one of the scientists that developed the V-2, also worked to produce the *Saturn* rockets that would eventually take US astronauts to the moon in 1969.[3] Without this capability to escape Earth's orbit, ironically funded in part by Hitler, the Apollo moon mission would not have been possible. Technology is the practical application of scientific and engineering understanding to solve a problem.

Relative technological capability measures the disparity between nations or organizations. In this sense, technological advantage over an opponent is essential. Both militaries and commerce benefit significantly from superior technology.

The military benefits because technology is 'superior' when the problem it has 'solved' is hostile military capabilities. The knight and the samurai both embodied the pinnacle of military technology through their combination of armor and blades. Firearms eliminated their technological advantage, rendering a peasant with a gun able to kill a warrior that had trained his whole life. Likewise, radar-guided surface to air missiles were bested by stealth technology in aircraft. Militaries are in constant competition with other militaries to ensure they have the best possible technology to solve their currently assessed problems.

Commerce also benefits significantly from technology but in a much more subtle fashion. While military superiority leaves flaming wrecks or a deterred nation, commercial superiority leads to profit and influence. Commercial technology is 'superior' if it solves a problem that previous technology did not solve as well or as efficiently. For instance, the telegraph replaced messengers; this radically increased communication speed. Those who controlled the telegraph became wealthy because they dominated the communication market. Likewise, cellphones solve the problem of communication instantly, and it is entirely possible to call someone on the other side of the world, and even use live translation services to circumvent language barriers.

Technology in astronautics is, like these other examples, both absolute and relative. Current problems that need to be solved are in-space construction using *in situ* resources, servicing satellites in orbit, and eventually space settlement construction - amongst others. Relative technological superiority in these areas will fuel national dominance in these industries, just as the development of computers and cellphones helped fuel the American economy for the last three decades.

Technology drives industry, which is the efficient development of technology into products for consumption or use by a nation or group. It includes the requirements to produce items but does not cover personnel. Limited industry in space is due to the currently unfavorable cost to benefit ratio, relative to Earth-based options. In other words, right now it is possible to accomplish most goals on Earth cheaper than in space. Economics function in the context of supply, demand, and investment; with increased demand and the investment that enables further production of supply, the industry will expand if it assesses that demand will stay constant. Since the US government has historically been the only customer for most space-related products, the demand has been low, and investment in improving capacity has also been inadequate. In this environment, relatively few companies have been able to survive because few commercial applications had been found for the space domain. The ones so far have been related to information and signals, where space has SATCOM-phones, Earth-imaging satellites, Direct TV, XM radio and Iridium as some examples. SpaceX, Blue Origin, and others are changing the playing field, increasing both supply and attempting to increase demand by providing new services from space. This increased demand will likely continue as SpaceX looks to expand the internet around the globe from orbit.

Industrial capacity for space has traditionally grown slowly and cautiously. This is partially because of the relatively small customer base and high cost of building and launching from Earth. There are predictions, however, that it will grow exponentially in the near future – fast enough to outstrip all US energy capacity in just 11 years.[4] This will be because of the sharply decreased launch costs driven by *in situ* manufacturing and significantly increased customer base that it will draw. Once the space industry can support itself with space-mined and space-manufactured capabilities, it has all the tools to continue self-replicating in space. This virtually eliminates the cost of launch from the equation entirely, leading to the possibility of exponential growth unfettered by terrestrial concerns.[5]

Personnel is a part of all aspects of astronautics, and, if and until AI can fully take over for humans in producing and operating space equipment, they will remain central. This category covers the skills needed to produce precision equipment made in clean rooms to the skills to operate them with both effectiveness and audacity. The requirement for audacity means that humanity will likely retain the edge for some time due to the human ability to innovate and use creativity to solve problems presented by the enemy and the environment. Personnel take a long time to build and grow. The difference in performance between a hundred fresh students and a hundred skilled operators will be significant because of the accumulated knowledge and experience of the latter group. Competitive astronautics requires a large and diverse group that can problem-solve collaboratively with different perspectives.

Once a technology has been developed, the industry has produced it, and personnel have learned how to operate it, commerce will attempt to make money off it. Commerce is like industry, both attempt to solve a market problem that will result in profits for the company. Industry produces items meant to fulfill a need. Commerce moves items from high-supply or low-demand areas to other areas that are either low-supply or high-demand. These items can also be natural resources, as exemplified by the current oil market and likely the future Helium 3 (He^3) energy market. He^3, when acquired in sufficient quantity, will revolutionize both terrestrial energy production and spaceflight due to the incredible energy it releases during nuclear fusion. Today, oil in Saudi Arabia is not scarce, and so it is shipped to other places to make profit; likewise,

He³ is plentiful on Uranus but incredibly rare, and therefore valuable, on Earth and the Moon.

Commerce is ultimately what will drive space development, pushing a nation from *exploration* and *expansion* solidly into *exploitation* because enterprising people want to gain personal wealth from what they find there. Further investment will follow successful missions and cause a mushrooming of growth that can far outstrip the most massive government attempt to compete.

The final element of astronautics is basing, which includes Earth-side launch bases, Lunar and other celestial settlements, and in-space stations such as O'Neill cylinders. For those unfamiliar with O'Neill cylinders, they are constructions kilometers long that spin to create false gravity through inertial force. Figure 3 shows an artist concept for the outside of two, while Figure 4 shows what the inside of one would look like. Once built, humans can live inside them and generate all their own

Figure 3 - Artist's concept for O'Neill Cylinders (credit NASA Ames Research Center)

necessary food, water, and oxygen to sustain life.[6] These can serve as ports and settlements and are used for the launch of spacecraft as well as resupply, and repair. They also serve as a safe destination for the ship.

Figure 4 - Artist's concept of inside of an O'Neill Cylinder (credit NASA Ames Research Center)

Earth bases are of differing utility based on the direction of a rocket launch and how close to the equator it is. Rockets launched from the most beneficial spots on Earth use propellant approximately 17% more efficient due to Earth imparting more of its inertia to the launching rocket.[7] Therefore, optimal launch bases on Earth are in limited supply and, in a real 'space boom' may become a valuable commodity. A limiting factor of this benefit is the cost of infrastructure required to transport spacecraft to the base. These costs can be high if the base is isolated or in a dangerous area. Some nations are just not well-positioned for space launch, so allied cooperation will likely be crucial to avoid prohibitive costs of sustaining the launch facilities.

Lunar and other bases serve in the context of "colonies" from Mahanian theory and serve as destinations and re-launch points to be able to move goods and personnel. The problem up to this point has been that, despite the desire to *expand* beyond Earth's orbit, it has not been economically viable. Instead, nations continue to compete inside Earth's gravity well. This appears to be changing, and space *expansion* is on the horizon. With the drive to expand humanity beyond Earth and the significant benefit of a rocket launch infrastructure on the Moon, Lunar and other settlements will serve as enabling bases to spread the reach of Astronautics.

Finally, proposed around 1970, O'Neill cylinders and other artificial space islands would also serve as in-space bases. These are particularly interesting because they can be placed on remote but strategic locations, like the Lagrange points, and have no intrinsic gravity. Instead of gravity, they rotate to simulate it through inertial force. This combination makes them an excellent option to serve as forts or production hubs. They can be produced in another location and then either move or be towed to the desired location, instantly fortifying it. If used as a production facility, it can create spaceships far more substantial than anything produced on a planet because ships will not have to escape any gravity well to reach space. Much like a whale will die if beached on land, a spaceship of this scale would likely be helpless once landed on a planet. Agile and free in the water, the whale's weight crushes it on land.

Technology must solve the issues that arise from space situations and, if competing, it must be superior to that possessed by hostile powers. The industry must be capable of producing systems in strategically impactful numbers to use the technology meaningfully. Personnel must command the produced systems well to prevent hostile tactics or audacity from overcoming inferiority in other areas. Finally, commerce drives further *expansion* and *exploitation* using trade. Bases spread the reach and allow a further toehold, both for production and resupply.

In a perfect world, astronautics would be the summation of space theory. However, the world is full of groups of people seeking benefits to the detriment of their fellows. Whether this is pirates, criminals, or hostile powers, there will always be a need for enforcement and protection in the face of these dangers.

Spacepower serves the dual purposes of securing the gains and capability that astronautics creates and projecting power to disrupt a hostile power's gains and capability. Space superiority will operate similarly to seapower in that, just as the sea serves as a highway, spacepower will form a valuable line of communication. England and America executed effective seapower strategies and forged enduring power through the combination of military and economic might. Lines of communication and bases helped secure gains and capability in both cases.

Even though the ocean is trackless and vast, it has lines of communication for commercial vessels due to natural movements of the water and wind. The discovery of the trade winds and the currents that run through the Atlantic created significantly more efficient means of travel for ships. Ships using these made the trips significantly faster, so they could make more trips and, therefore, generate more money for their owners. This discovery meant that all ships desired to use these – yielding an established line of communication in the trackless sea.

Narrows, straits, and canals also force shipping through a single location, creating a chokepoint. Terrestrially, new critical points are largely natural, but some are man-made, with the Suez and Panama Canals serving as two specific examples. These massive construction projects served to significantly increase the power of nations that controlled them through both military and commercial applications.

In space, there are likely equivalencies to all of these. Gravity wells, solar winds, and orbits all create some areas more efficient to move than others. Lagrange points are areas of stable gravity, and solar winds drive particles throughout the solar system. Orbits of planets provide potential kinetic energy to speed ships on their way. While these phenomena have been heavily studied, humanity sailed the ocean for centuries before they discovered some features of the waves. Similar breakthroughs are likely awaiting discovery in space.

The other element that spacepower must project is capability from ports. Ports are where other domains convert resources into power that can then be projected into space – for example, land-produced satellites are sent into space. In a naval example, a ship constructed on land is put into the water, projecting land-produced power into the naval domain. It is also where spacepower already in the domain can refit and resupply. Whether termed ports, bases, or settlements, the primary purpose is the same.

If the location to be protected is on an island or some other isolated body, seapower has proved successful in defending against significantly larger land-based forces. England ruled the waves for hundreds of years, and their island has not been invaded successfully since 1066. Seapower can serve as the shield, deterring attempts to engage at all, or if engaged, can eliminate the enemy fleet.

However, if the capability is on a shared continent or some other contested landmass, then seapower can serve as a line of communication to the coastal base even if it is surrounded or sieged. Because it is never truly isolated, it can survive and receive reinforcements via the sea if appropriately fortified. Seapower can serve as both shield and pack mule.

Spacepower can serve much the same role but has a significant benefit that seapower lacks – reach. Seapower will always have this limitation when dealing with land-based powers because, while Earth is two-thirds water, the remaining third requires land power to influence it.

Missiles and other weaponry can reach inland targets. However, they are limited in size and efficacy and must survive the nation's countermeasures. Spacepower does not have this limitation, as it spans the entire Earth and its assets can move far more rapidly than even airpower. In orbit, it can act like aircraft, able to strike anywhere it desires but with many of the drawbacks of airpower as well. Spacepower operating in this context is held at risk from ASATs launched from below, co-orbital threats from other satellites and possible defensive measures from extra-orbital objects. If spacepower is tied to the celestial island, it will always be vulnerable to these because it operates inside their umbrella – not dissimilar to naval forces operating in the littorals, threatened by hostile vessels and missiles from the shore. Especially if the space around Earth were explicitly militarized, spacepower could only operate in that vicinity at its own peril.

This is the reason that modern spacepower's fixation on survivability and resilience in the littoral is ultimately Sisyphean. [8] Make large systems more survivable, the enemy makes its systems more lethal. Multiply systems for resilient capabilities and the enemy matches your numbers. The problem is that holding the littoral indefinitely is virtually impossible due to the predictability of your own systems, the number of vectors that the threat can approach from, the speed they can do it, and the cost of hardening systems to completely prevent it.

For a similar example, the US did not radiation-proof all its cities during the Cold War to protect them from the USSR's nuclear strike. Instead, it occupied a position of strength that threatened annihilation to the USSR if it decided to fire. A similar situation presented to an adversary, the destruction of their satellite fleets and subsequent strategic defeat, will likely have the same effect of deterrence at significantly less cost than exquisitely engineering all satellites to meet all possible scenarios.

Likewise, blue water navies do not seek to fight in the littorals, but rather have dedicated craft that can perform well there and support them with the true power of the fleet, the blue water capability that is relatively immune to terrestrial counter-punches. The recommended option for spacepower would be to embark on a similar strategy with space-based capabilities. Build the blue water capability and give it the ability to project into the littorals for a short time as needed. This reduces vulnerability of the systems as blue water could project anywhere in the world and not have to sit in orbit for years while the enemy can evaluate and target at its leisure. Littoral spacepower without blue water will always be vulnerable.

Once it projects out of orbit into the silent sea, spacepower operates more like traditional seapower, serving to connect planets and bases. It also greatly minimizes the previous weaknesses identified – satellites in orbit are of little threat to a vessel orbiting the moon, and ASATs from Earth would need to be massive to have any hope of successfully targeting it. Once away from the cluttered and congested planetary littorals, spacepower unfurls its sails as the master of the silent domain. The only threats to spacepower at that point are hostile spaceships or natural phenomena.

Before this future can be achieved, both astronautics and spacepower must reach fruition. Currently, astronautics has matured enough to provide the impetus for the development of true spacepower, which is leading spacepower out of its infancy. As astronautics continues to develop, national spacepower will have to develop as well to keep pace and protect the fledgling market. Those nations that do not will find their astronautics stunted or co-opted by other powers that are willing to invest in the development of spacepower. Nevertheless, astronautics must lead the development of spacepower if spacepower is to be enduring.

―――――――――――

[1] Brent Ziarnick, *Developing National Power in Space: A Theoretical Model*. (North Carolina, McFarland & Company, Inc. Publishers, 2015.), 15-28.
[2] Rob Garner, "Dr. Robert H. Goddard, American Rocketry Pioneer," National Aeronautics and Space Administration, edited 3 August 2017. Accessed 30 April, 2020: https://www.nasa.gov/centers/goddard/about/history/dr_goddard.html
[3] Alexis C. Madrigal, "Remembering the Nazi Scientist Who Built the Rockets for Apollo." (The Atlantic, 23 March 2012). Accessed 30 April 2020: https://www.theatlantic.com/technology/archive/2012/03/remembering-the-nazi-scientist-who-built-the-rockets-for-apollo/254987/
[4] Brian Wang, "Exponential industrialization of space is more important than combat lasers and hypersonic aircraft," Next Big Future, 26 October 2017. Accessed 11 June 2020: https://www.nextbigfuture.com/2017/10/exponential-industrialization-of-space-is-more-important-than-combat-lasers-and-hypersonic-fighters.html
[5] Wang, "Exponential industrialization."
[6] O'Neill, *The High Frontier*, 79.
[7] Everett C. Dolman, *Astropolitik: Classical Geopolitics in the Space Age*. (Frank Cass Publishers, New York, New York, 2002), 77-78.
[8] Sisyphean Labor refers to the Greek mythology of Sisyphus, who the gods made to roll a giant boulder up a hill as punishment. Whenever he would get it to the top, it would roll down again. A hopelessly cyclical task; https://www.thefreedictionary.com/Sisyphean+task

Chapter 4
Chinese Spacepower Theory

For China, space is the key to achieve their long-term ambition of becoming a primary power in the international system...China's space-power theory is undergirded by its expansionist view of space as a geography, and a means to achieve global legitimate leadership through the economic exploitation of the inner solar system.

--Dr. Namrata Goswami

Discusses Chinese theory on space and some of its strategic underpinnings. China plays 'Go,' not Chess – a game of maneuver, terrain occupation, and influence rather than direct conflict. China's theory does not foresee an interstellar war to secure dominance, but rather securing the superior position in space so they never have to.

Chinese spacepower theory can be boiled down into three words: Geography, Legitimacy, and Economy.[1] They are the three layers of meaning in the Chinese concept of space development. First, and the grounding principle, is the view of space as geography, an area that can be owned and controlled. Second, legitimacy refers to the view of space as a forum for leadership and international affirmation of the regime. Third, and most recent to develop, is the economic aspects of space and the ability for that to bolster their domestic economy. This is captured in Figure 5, which shows the building of Chinese space theory as a triangle.

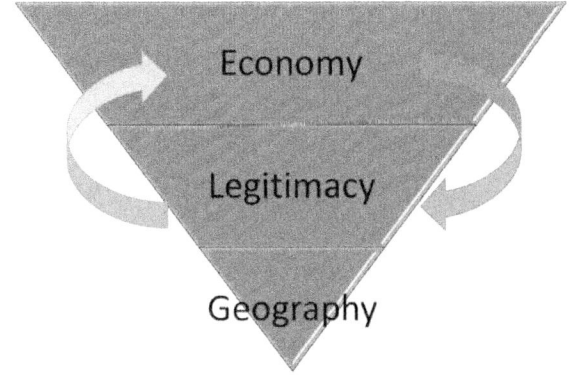

Figure 5 - Chinese Space Theory (original by author)

    The Chinese understanding of space as an extension to geography is best understood in context of the realist school of international relations.[2] Simply put, power grows through the acquisitions of territory and resources for the state, and states compete with each other in an anarchical system. Therefore, states must continuously attempt to expand their power to retain maximal relative advantage compared to other countries in the system; defeat for one usually equates to victory for another.[3] Also intrinsic to this belief is that information and exploration, as well as naming, are crucial to claiming ownership.[4] If 'ownership is 9/10 of the law,' and property follows a progression of *exploration* and *exploitation,* then the concept follows that the Chinese will try to progress as far down that path as possible. The consequence of this first principle is that the Chinese believe that ownership is at stake and that they must encourage the growth of Chinese holdings while reducing those not aligned with Chinese interests. The closest analogy is *Go*, not Chess – strategic maneuver to gain positional advantage can ensure continued dominance without fighting wars.

Second, space is a venue for the legitimacy of the regime for two overarching reasons.[5] Primarily, being able to discover and name space locations projects the image of leading the world in space technology. Secondarily, developing space access required for that *exploration* allows for sharing or denying that too other nations.

Discovery and cataloging of space objects bolsters claim of national achievement, and therefore legitimacy of the regime that has overseen those accomplishments. China has engaged in this on the Moon, naming locations they have occupied as, "Statio Tianhe", "Mons Tai", and "Guang Han Gong."[6] More than just naming them on internal documents, China has had those names approved by the International Astronautical Union, conferring international legitimacy on their 'first presence' claim. This is an example of Chinese diplomacy supporting their space program.[7] Naming implies ownership. Cities are named by their founders and may be renamed if ownership changes. Parents name their children. Inventors name their work. Nations name the areas they control. This is one major way China enhances its legitimacy through space, using it to 'prove' their superiority as a nation by pushing the boundaries of *exploration.*

Also, because space access and *exploration* are cost intensive. Therefore, smaller nations' access to orbits and crewed platforms has required patronage by countries that can support the industry and architecture. Nations tend to align to a nation that can provide access. The goal of these interactions is not cooperation for the betterment of all but increasing power through income and bolstering international support for China. Countries in Africa and Europe have space programs that China is more than willing to host on their rockets.[8] This extends to the UN as well, where China has already offered to host other nation's science projects on the Chinese Space Station when it is constructed in 2022.[9] Both of these actions seek to collect friendly nations that owe China a debt for their assistance. Because power distribution in the system is vital, China is weakening the US monopoly on astronautics and spacepower even as they use other nations business to build their own. In China's perspective, the current world order is not democratic but hegemonic, with the US at the helm. China's answer is to eventually use space leadership through Moon settlements, space-based solar power, and asteroid mining, to set up a new system that reduces the position of the US and, in providing benefit for the rest of the world, gain legitimacy as the new ascendant hegemon.[10]

Third, and most recent to develop, is the Chinese concept of space as a domain for economic benefit.[11] There are financial elements of interest in the principles above, but those are a means and not ends. In this most recent development of China's space theory, economic development serves as an end. The massive economy of China has served as a source of national pride and a tool to bludgeon those opposed to Chinese terrestrial interests for nearly a decade. In exploiting space for the resources and production capability, China hopes to continue this evolution and propel its economy past terrestrially based markets. Those nations focused on Earth cannot hope to compete with the scope of resources, area, or propellant-efficiency that space economic development offers. On the Moon alone, there is water, titanium, and He^3.[12] Water is crucial for propellant and life support, allowing for permanent human occupation of lunar bases. Titanium is a valuable metal used in high-temperature and high strain applications. He^3 is the resource of the future, with the potential to power very clean-burning nuclear fusion reactors that could revolutionize both the earth's electrical power and space flight. Whomever controls the lion's share of these resources will reap the most economic rewards for their country. These will increase absolute power and, as discussed in the first principle, increase China's relative power and clout in the world system.

Finally, there is a noted interaction between the three levels as well. Conceptualizing space as a geography provides for ownership, which can be gained and lost. It is this concept of ownership that enables the next step of Chinese space theory, which is legitimacy. Legitimacy flows from both the Chinese Government's ability to declare intentions and execute them, as well as having authenticity that comes from space ownership. Kings without kingdoms, or land, are generally not kings long. Finally, viewing space as a bolster to the economy is crucial to China because it has a symbiotic relationship between its economic power and the ruling party's legitimacy. When it is doing well, the party's rule is legitimate. Knowing this, the party does everything to encourage economic growth. Therefore, legitimacy and economy feed off each other in China's current system – and space is no different. China has tied legitimacy in space to its ability to take and hold terrain as well as to reap the economic benefits from the resources found there. If China were not able to hold terrain or exploit resources, it would remove a significant pillar of legitimacy for the Chinese Communist Party.

Chinese space theory is comprised of three concepts – geography, legitimacy, and economy. These are synergistic with each other, creating a holistic and expansive vision of space. As China prepares to build a station on the Moon in the next twenty years, the world needs to know why they are doing it. China is doing more than *exploring*. They do *explore*, but their theory embraces *expansion* and *exploitation* wholeheartedly. Failing to understand this, and contest China's occupation, will result in China being able to *exclude* actors they consider hostile from the Moon and other bodies, with very little able to be done in the domain to challenge it.

---

[1] Namrata Goswami (Minerva Grantee and Senior Analyst with Wikistrat), interview by the author, 15 January 2020.
[2] Goswami, 15 January 2020.
[3] Jervis, Robert. "Realism, Neoliberalism, and Cooperation: Understanding the Debate." *International Security* 24, no. 1 (1999): 42-63. Accessed March 16, 2020. www.jstor.org/stable/2539347.
[4] Jason Smith, *To Master the Boundless Sea* (Chapel Hill, University of North Carolina Press, 2018), 5-7.
[5] Namrata Goswami (Minerva Grantee and Senior Analyst with Wikistrat), interview by the author, 15 January 2020.
[6] Namrata Goswami, "Explaining China's space ambitions and goals through the lens of strategic culture," The Space Review, 18 May 2020. Accessed 17 June 2020: https://www.thespacereview.com/article/3944/1
[7] Goswami, "China's space ambitions"
[8] Andrew Jones, "Long March Rocket Launches Chinese-Brazilian Mission and 1st Ethiopian Satellite", Space.com, 21 December 2019. Accessed 21 May 2020: https://www.space.com/china-long-march-4b-rocket-launches-9-satellites.html.

[9] "United Nations/China Joint Announcement Event on Selected Experiments on board the China Space Station," United Nations, Vienna, 29 May 2019. Accessed 17 June 2020: https://spacepolicyonline.com/events/un-china-announcement-of-experiments-for-chinese-space-station-june-12-2019-vienna-austria/

[10] Senate, *China in Space: A Strategic Competition? Hearing before the U.S.-China Economic and Security Review Commission*, 116th Cong., 1st sess., 2019, 82-85.

[11] Namrata Goswami (Minerva Grantee and Senior Analyst with Wikistrat), interview by the author, 15 January 2020.

[12] Paul Spudis, *The Value of the Moon: How to Explore, Live, and Prosper in Space Using the Moon's Resources* (Washington, DC, Smithsonian Books, 2016), 119-124.

For further reading:
Qiao Liang and Wang Xiangsui, *Unrestricted Warfare: China's Master Plan to Destroy America*, Echo Point Books & Media, Reprint Edition, 10 November 2015.
Liu Mingfu, *The China Dream: Great Power Thinking and Strategic Posture in the Post-American Era,* CN Times Books, Inc, New York, NY, 2015.

Chapter 5
American Spacepower Theory Today

We do not want a conflict to begin or extend into space. We want to deter that conflict from happening. The best way I know how to do that is from a position of strength.

--Gen. John "Jay" Raymond

In contrast to China, the US has a wide range of theories but the military – who currently drive space theory – primarily envision it as a domain to support the warfighter on the ground. It looks inward because that is how funding is garnered from Earth-based governments. It does not see outward, looking to the future.

There are many spacepower theories in the United States, many built on the seapower theory of Alfred Thayer Mahan. These lie on a spectrum, ranging from extremely militarily aggressive to much more commercially focused.[1] What all of them have in common are the themes of control, expansion, and development. Most American space theories have an element of space control in them – to be able to use space and deny it to an adversary in the case of war. Many of them also address *expansion* – going beyond the celestial island of Earth into the tremendous silent sea beyond and developing bases and settlements. These are not far divergent from the Chinese space theory, but the problem is that they have not been widely read even inside the US military space community. This ignorance of theory has detrimental effects on both the understanding of the strategic situation and the ability to craft an appropriate strategy, using military means for political ends.

The earliest space theory writing comes from a book published in 1958, *Spacepower: What It Means To You*. It was written the year after Sputnik shocked the Western world with a decisive display of Soviet rocket capability.[2] The book was far ahead of its time, already drawing distinctions like *contiguous* space flight, which was around the Earth, and contrasting it with *interplanetary* space flights to other bodies in the Solar System. It went so far as to anticipate *intergalactic* flights between the Milky Way galaxy and other galaxies.[3] Lacking is the presence of the naval undertone commonly found in modern space theories. However, for that, it did have an expansive, but more general, view of space, taking account of moral, spiritual, and psychological reasons for exploration, in addition to the economic, scientific, and military.[4]

While there are too many space theorists to discuss in entirety, there are three major ones that can serve as archetypes – the geopolitical approach of Everett Dolman, the very Mahanian approach of Brent Ziarnick, and the Earth-focused approach of Bradley Townsend. Dolman's approach is focused on the importance of space in a military context. In contrast, Ziarnick took a more system-approach and focused on the production and projection of national power into space. Townsend does not highlight space as an independently valuable domain, instead seeing it as information driven and terrestrially tied.[5]

Dolman's *Astropolitik* explores space from the perspective of national conflict and power projection. His book's perspective serves as a bookend for the most extreme warmonger view possible of space. Space capability was birthed by combining several technologies produced during World War II and fueled by the ongoing competition between the US and the Soviets during the Cold War.[6] His book explores what the ultimate expression of pure spacepower would look like. His dramatic suggestion was to have the US withdraw from all outer-space treaties and declare it would embark on the "total domination of space."[7] When written in 2002, this was indeed a possibility. There was no other nation to challenge the US in space. The Soviet Union had recently collapsed, China was still ascending, and the Iranian space program was inconsequential. That is not the situation anymore, with the Chinese space program a significant threat to the US in space. However, his assessment of the US current space theory application by leadership appears to be accurate when he describes:

> Current US space strategy is focused more on technological capabilities, and to a lesser extent in developing military and commercial capabilities. Given ambiguous and weak leadership from the top, strategy is perhaps naturally timid, hedging, elusive, evasive, and contradictory. When vision is not provided, followers will focus on the concrete: current and future technology and system applications.[8]

Dolman proposes an unacceptable course of action, essentially attempting to dominate space through a pure military regime designed to eliminate the possibility of other nations *expanding* into space. Putting that aside, his theory's most valuable contribution is putting the value of space in context. In the context of *Geopolitik* theory, the central landmass of Eurasia is seen as most valuable to possess due to its inherent ability to dominate the rest of the world through superior production capability and resources. Likewise, *Astropolitik* argues that space is Eurasia and Earth is an island to be dominated. As discoveries of more resources in space continue and industry gets closer to being able to *exploit* them, this predicted future is inching closer to reality.

On the opposite side of the military/civilian involvement spectrum, Ziarnick's *Developing National Power in Space* focuses almost exclusively on the process of how a nation should go about developing astronautics. He envisioned the development of astronautics and spacepower as two deltas, or triangles, which he described as 'grammar' and 'logic' – an ode to Clausewitz's description of war.[9] The grammar, or production, triangle flows from the basis of production, shipping, and settlements to produce spacepower and access.[10] This power is transferred to the national will and directed down into the three spokes of the logic, or usage, triangle – which is comprised of economic, political, and military applications. [11] Figure 6 captures and the interactions between them.

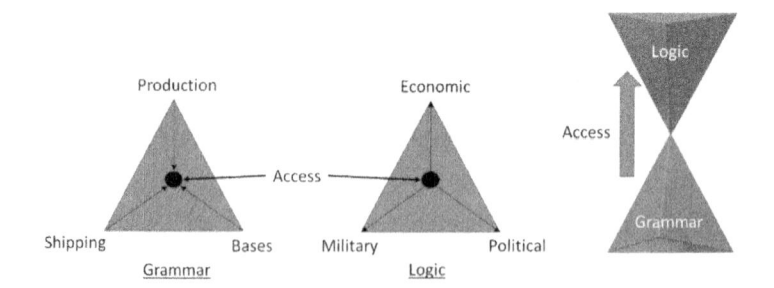

Figure 6 - Triangles of Grammar and Logic with interrelations, summary of Brent Ziarnick's theory of Developing National Power in Space (original by author)

This theory does an excellent job of embracing the holistic nature of astronautics and highlight both the crucial base that grammar provides and the actual applications of space capability. Of note, nowhere in his logical application delta does science appear; it is not a valid end for space development. A step and a means, yes, but not an end. There is no reason for a nation to spend billions of dollars developing space technology and then expend production to go to space for the weak reason of "scientific curiosity." If there is no application or benefit to be gained by the nation, it is a waste of the nation's resources.

Taking a very Earth-centric but more civilian/military balanced approach, Bradley Townsend addresses some who believe current conditions do not justify the Space Force as an independent service.[12] Following the tendency of most authors, he conflates spacepower and astronautics, lumping them all under his term, *space power*, "which includes the commercial and political aspects."[13] This perspective does not allow for a difference between civilian satellites, which one of the authors he cites acknowledges can be bought by the highest bidder, and military assets that belong solely to a nation-state.[14] It also does not differentiate between the purpose of the satellite in question, whether to primarily make a profit or project national power. Satellites that deal in information for economic gain are different from, for instance, the referenced Voice of America satellite whose sole job is to broadcast American messages into hostile countries that restrict information from their people, like Iran.[15] One operates for the highest bidder and can, therefore, have its services bought and the other is a national asset that does not need to be economically viable to continue functioning.

The crux of Townsend's argument about spacepower's purpose is encapsulated in his statement that, "[s]pace is an enabling domain for terrestrial military operations. Actions that take place in space are relevant only to the degree that they impact events on Earth."[16] This perspective is unduly-focused on terrestrial concerns to the exclusion of all other elements. It fails to look with sufficient scope – in particular, it is very military in how it defines "impact." In context, replacing "impact" with "informationally influenced" would make sense. That definition excludes other capabilities of space, particularly the economic. As noted earlier, China is planning on creating a cislunar economic zone, which will have direct impacts on both information and economic flow on Earth. While that is not explicitly the purview of the USSF alone, it is something that any spacepower theory must consider. Economic dominance by China in space would negatively 'impact' the US' calculus of conflict on Earth. However, his current definition does not appear to incorporate that possibility.

The major issue with the premise of Townsend's argument is found earlier in his article, where he says, "Developing a theory of space power is made doubly difficult because, unlike the sea, space is an untested domain. Humanity lacks any empirical evidence on the nature of conflict within it."[17] There are two main flaws to take issue with – the assumed inhumanity of combat and the narrow spectrum of evidence allowed.

First, by inhuman combat, Townsend seems to see space as possessing a different nature of conflict than air, ground, or sea. This is a flaw; the nature of conflict does not change between domains because humanity remains the consistent factor throughout. Some details of combat will change due to domain-specific elements like microgravity and inertia in space, but the essential element of humanity will not change. Therefore, space will be subject to conflict in a similar fashion to these traditional domains. This same underlying nature is why SDT, proposed in this book, uses the template of other domains to attempt to gain as accurate an assessment as possible for future space conflict.

Second, even as he acknowledges gratefully that there has never been a war in space, he makes suggests it is a requirement. It is undoubtedly unique in the history of warfare to project forward to anticipate combat in a domain not yet blooded, but it is still possible through analogy. Reasoning by analogy is an accepted method of seeking the truth about an unknown thing – comparing a known and an unknown to attempt to explain the unknown.[18] There has never been space warfare, but there has been warfare with similar characteristics in other domains. By adapting these, and taking the domain-unique characteristics into account, a rough picture is certainly possible. Taking all of this into account, analysis by analogy is the most appropriate path to develop space theory. To wait for a conflict to occur before developing a theory is to court defeat by an enemy able to pre-reason a better theory and strategy.

The issue with this backward-looking is that it fails to anticipate the future. Crafting a system of thought based solely on empirical evidence is history, not theory. A theory must look toward the future, and grapple with the enduring. Taking the enduring as a guide and looking to the possible, a strategy takes theory and applies it to national goals desired in the current situation. Random technical innovation provides valuable assistance to achieve these goals but might not produce the required machinery on its own. Strategy must drive technological development and not be driven by it.

While thinkers have produced several notable spacepower theories in the US, the USSF, and its predecessor, USAF Space Command, has explicitly abstained from embracing any of them. With the formation of the USSF, this is beginning to change. The USSF will soon release the Space Power Capstone to clarify the USSF's perspective, which is commendable considering how soon the new service is achieving this milestone.

In short, the US has developed many spacepower theories but embraced none of them. Because of this, there has been no declared space plan. In its place is the automatic strategy created by organizational decision making, continuing to produce military spacepower as if the Cold War were still on with the USSR. This lack of theory runs the risk of fulfilling the quote at the beginning of one of the first spacepower books. It quotes the Bible, "Where there is no vision, the people perish."[19] SDT, fully outlined in chapter 12, attempts to fill this gap and provide a universal approach to dealing with space and assist in building a roadmap to the stars.

---

[1] Russell Rumbaugh, "Six Competing Visions for a Space Force," War on the Rocks, 7 October 2019. Accessed 11 June 2020: https://warontherocks.com/2019/10/six-competing-visions-for-a-space-force/.
[2] Donald Cox and Michael Stoiko, *Spacepower: What It Means To You* (Philadelphia: The John C. Winston Company, 1958).
[3] Cox and Stoiko, *Spacepower*, xvii.
[4] Cox and Stoiko, 3.
[5] Townsend, 20.
[6] Dolman, 89-91.
[7] *Ibid*, 156.
[8] *Ibid*, 155.
[9] Ziarnick, Grammar and Logic Triangles and Clausewitz, "War has a grammar and logic"
[10] Brent Ziarnick, *Developing National Power in Space: A Theoretical Model* (Jefferson, NC: McFarland & Company Inc., 2015), 16.
[11] Ziarnick, 25.
[12] Townsend, 11.
[13] *Ibid*, 13.
[14] *Ibid*, 17-18.
[15] *Ibid*, 15.
[16] *Ibid*, 20.
[17] *Ibid, 14.*

[18] Paul Bartha, "Analogy and Analogical Reasoning," Stanford Encyclopedia of Philosophy, 25 January 2019. Accessed 21 May 2020: https://plato.stanford.edu/entries/reasoning-analogy/.
[19] Cox and Stoiko, i.

For more reading on the Blue Water school of space theory, see:
- Simon Worden and John Shaw, *Whither Space Power? Forging a Strategy for the New Century,* Air University Press, Maxwell Air Force Base, AL, September 2002.
- Stefan Possony, Jerry Pournelle, and Francis Kane, "The Strategy of Technology," jerrypournelle.com, 1997. Accessed 11 June 2020: https://www.jerrypournelle.com/slowchange/Strat.html
- Daniel Orrin Graham, *High Frontier: A New National Strategy*, High Frontier, 1982.
- Mark Stokes, Gabriel Alvarado, Emily Weinstein, and Ian Easton, *China's Space and Counterspace Capabilities and Activities,* US-China Economic and Security Review Commission, 30 March 2020. Accessed 11 June 2020: https://www.uscc.gov/sites/default/files/2020-05/China_Space_and_Counterspace_Activities.pdf.
- Lamont Colucci, "The Space Review: A Space Service in support of American grand strategy," The Space Review, 25 February 2019. Accessed 11 June 2020: https://www.thespacereview.com/article/3664/1.
- Lamont Colucci, "Why It's Time for a 'Triplanetary' Economy," Newsmax.com, 31 January 2020. Accessed 11 June 2020: https://www.newsmax.com/lamontcolucci/space-mars-earth-moon/2020/01/31/id/952148/.
- Michelle Shevin-Coetzee and Jerry Hendrix, "From Blue to Black: Applying the Concepts of Sea Power to the Ocean of Space," Center for a New American Security, 18 November 2016. Accessed 11 June

2020: https://www.cnas.org/publications/reports/from-blue-to-black.
- *Toward a Theory of Spacepower: Selected Essays*, edited by Charles Lutes and Peter Hays, Institute for National Strategic Studies, National Defense University, 1 February 2011. (Chapter 8, Dennis Wingo) Accessed 11 June 2020: https://ndupress.ndu.edu/Portals/68/Documents/Books/spacepower.pdf.
- Mir Sadat's testimony, Transcript: US Space Strategy and Indo-Pacific Cooperation, chaired by Patrick Cronin, Musashi Murano, and H.R. McMaster, Hudson Institute, 13 November 2019. Accessed 11 June 2020: https://s3.amazonaws.com/media.hudson.org/Transcript_%20US%20Space%20Strategy%20and%20Indo-Pacific%20Cooperation.pdf.
- Newt Gingrich, H.R. 4286, 97th Congress, 1st Session, 28 July 1981. Accessed 11 June 2020: https://www.congress.gov/bill/97th-congress/house-bill/4286?q=%7B%22search%22%3A%5B%22H.R.+4286+97th+Congress%22%5D%7D&s=1&r=1
- Jim Bridenstine, "This is out Sputnik Moment," OKGrassroots.com, 6 November 2016. Accessed 11 June 2020: https://okgrassroots.com/?p=642815.
- Roger Lenard, "A treatise on the formation of a US Space Force," The Space Review, 22 January 2018. Accessed 11 June 2020: https://www.thespacereview.com/article/3411/1.
- Sean McClain, "Celestial Sentinels: A Framework for Cis-Lunar Space Domain Awareness in 2035," Thesis for Air Command and Staff College, Maxwell Air Force Base, AL, March 2020. Accessed 11 June 2020: https://www.linkedin.com/posts/namratagoswami_celestial-sentinels-by-sean-mcclain-activity-6664614183398629376-K-IH.

For reading that comes from "Military Space" or Brown Water space theory perspectives, see:
- Bleddyn E. Bowen, *War In Space: Strategy, Spacepower Geopolitics,* Edinburgh University Press, Edinburgh, UK, 2020.
- Brad Townsend, "Space Power and the Foundations of an Independent Space Force," Air & Space Power Journal, Winter 2019.

For an idealistic perspective on space:
- Clay Moltz, *The Politics of Space Security: Strategic Restraint and the Pursuit of National Interest,* Stanford University Press, Stanford, CA, 2019.

For pro-sanctuary perspectives on space:
- Joan Johnson-Freese, *Space as a Strategic Asset,* Columbia University Press, New York, New York, 2007.
- Michael O'Hanlon, *Neither Star Wars Nor Sanctuary: Constructing the Military Uses of Space,* Bookings Institute Press, Washington DC, 2004.

For pro-weaponization perspectives on space:
- John Klein, *Space Warfare: Strategy, Principles and Policy,* Routledge, London, 2006.
- John Klein, *Understanding Space Strategy: The Art of War in Space,* Routledge, London, 2019.
- Peter Hays, "United States Military Space: Into the Twenty-First Century," INSS Occasional Paper 42, USAF Institute for National Security Studies, USAF Academy, Colorado, 2002.

For further reading on space:
- Joan Johnson-Freese, *Space Warfare in the 21st Century: Arming the Heavens,* Routledge, London, 2017.
- John Collins, "Military Space Forces: The Next Fifty Years," Congressional Research Service, 12

October 1989. Accessed 11 June 2020: https://www.spacefaringamerica.com/wp-content/uploads/2020/05/military_space_forces_collins_1989_csr_report.pdf

Chapter 6
Chinese Space Strategy

To be second to the Soviet Union in establishing an outpost on the moon would be disastrous to our nation's prestige and in turn to our democratic philosophy. Although it is contrary to United States policy, the Soviet Union in establishing the first permanent base, may claim the moon or critical areas thereof for its own. Then a subsequent attempt to establish an outpost by the United States might be considered and propagandized as a hostile act.

--Statement from the USAF's Project Horizon

China's space strategy does not have the excitement, at first glance, of a tiger stalking and charging its prey – bringing it down in a shower of mud and blood. It has chosen the path of the python, squeezing its prey silently until it asphyxiates. This strategy is integrated into their overall national strategy. Without control of space, China will stutter in its ascent.

China has a strategy, which is the planned method to apply resources to a problem and achieve the desired end state. The Chinese space program, put forward by various space agencies in China and the Chinese Communist Party itself, has an aggressive timeline. The schedule, as put out by the Chinese official science and space institutions, is best summarized in Namrata Goswami's congressional testimony:[1]

- 2020: First Mars Probe[2]
- 2022: Permanent Space Station, *Tianhe-1*[3]
- 2024: Chang'e 6 to bring back samples from the Lunar South Pole[4]
- 2025: Complete the first 100kW SBSP demonstration at LEO[5]
- 2029: Mission to capture a Near-Earth Asteroid (NEA)[6]
- 2030: Probe to Lunar South and North Pole[7]
- 2030: 1 MW of SBSP power demonstration in GEO[8]
- 2035: 100 MW SBSP will have electric power generating capacity[9]
- 2036: A crewed mission to moon and establishment of a lunar research base[10]
- 2040: Nuclear powered space fleet to be ready[11]
- 2045: Most advanced Space Nation[12]
- 2049: 100th-year celebration of the establishment of the People's Republic of China (PRC).[13]
- 2050: First commercial-level SBSP systems come online[14]

These milestones come from official statements, so China has a vested interest in ensuring accomplishment to save face. More than that, it also plays into China's legitimacy in their space theory. The party has based part of its right to rule on the ascendancy of its continued economic growth and glory, furthered by the space program.

One of the drivers of this strategy is the anticipated strategic shortfall in China's energy sector, which would threaten their continued economic strength. According to a report released by China's National Bureau of Statistics in February 2009, China's expected energy needs in 2050 will significantly outstrip its capability – by 10.5% even after allowing for growth in all other sources of energy.[15] This energy gap is what space-based solar power (SBSP) is targeted to fill, enabling continued growth and energy independence for a future China.[16]

By securing a Moonbase, developing SBSP, the use of in-space resources, and construction on-orbit, China could produce solar power satellites at a rate no other nation can hope to match because of the synergistic nature that astronautics and spacepower provide.[17] This production will let them dominate the energy market, which will subsidize further space expansion and growing their economic and political power in the world. Indeed, inside six years of full production, China could be completely energy-independent using SBSP satellites exclusively.[18] While a direct military attack conducted on these satellites would disrupt progress, it would have the double drawbacks of appearing highly aggressive and disrupting other nation's that are relying on China's SBSP satellites. It would not be realistic to attack these satellites without a declared war, and collateral damage would likely be significant.

An additional benefit of this strategy is the hegemony it can provide. Economic domination leads to hegemonic status if done in the proper order – which is to 1) dominate the production of the most valuable commodities, 2) dominate trade by becoming transporter of choice, and 3) use the profits from production and trade to become the financial or banking leader.[19] Once China begins to dominate the SBSP market, they can use that leverage with all other instruments of national power: diplomatic, informational, military, and economic. Diplomatically, they would control an easily redirected energy source that can be provided to anyone or taken away at will. Informationally, it feeds the narrative of China as a benevolent superpower, providing energy to everyone who needs it. Militarily, it increases the movement capability and deployment of their forces by providing energy automatically wherever they deploy. Additionally, if the solar arrays have dual-use systems capable of space-targeting, then their offensive space capabilities would also be significantly increased. This includes both space-to-space and space-to-ground capabilities. Finally, economically, China will be freed from importing energy and will become completely independent, while also creating a new industry, which they will dominate.

A second driver of this strategy is the Chinese understanding of history. During the Ming dynasty, in 1525 CE, the decision was made to burn the treasure ships of the Chinese admiral and explorer Zheng He.[20] This failure to exercise seapower and *expansion* is assessed by Chinese strategic thinkers to have resulted in the Chinese humiliation during the 19th century.[21] It is difficult to underscore the depth of suffering inflicted on the Chinese people by both Western powers and Japan leading up to World War II. The scars run deep, festering still to this day.[22] This emotional scarring provides grist for skillful politicians to cast a narrative of the need for Chinese expansion and modernization. China has determined that the strategic misstep of failing to look outward will not be repeated.[23] They are pursuing *expansion* of their space infrastructure and capability; they will not burn their ships a second time.

After examining some undercurrents that drive China's strategy, it is time to look at the overarching observations. First, China's strategy is an excellent example of 'indirect warfare' in the tradition of Sun Tzu. To be victorious, China does not need to have a destructive war against the world's remaining superpower or even engage in the kind of proxy warfare that was so prevalent in the Cold War. Instead, it simply needs to *expand* into space, which is currently largely uncontested, and begin to *exploit* the plentiful resources found there – material and energy.

If China risked a war, it could receive significant damage and escalate to nuclear exchange. A war between China and the US is in neither country's interest – and so China has chosen to avoid it altogether. While it has a military that is eclipsed by few in the world, it is not the military that will grant China the victory, but the economic power and energy production. By becoming the foremost SBSP energy producer and using that to further fuel their expansion in SBSP and the Moon, China can become dominant in space. Attempting to contest China militarily will not work because, once they have begun *exploitation*, their space industry will be able to outstrip any rivals – letting them rapidly field overwhelming numbers of space systems in any space war.[24] In short, if China can *expand* into space uncontested and achieve space industrial dominance, there is little hope of breaking their hold on the world order.

Second, while scientific exploration is a significant element of China's program and it will likely accomplish those objectives, it has a direction far beyond mere scientific curiosity – it bends sharply to *expansion* and, quickly, *exploitation.* The stated objectives of permanent and crewed bases on the Moon and other locations reflect a Chinese space program, half the age of its older rival, that is already striving to surpass the US. This speed is not because of some genetic advantage or technological breakthrough, but a strategic focus.

    China has a different space program, primarily economic and expansive, versus military, or scientific. It is not focused on racing the US to some milestone, like putting humans on Mars, but the domination of resources, territory, production, and delivery in and from space. This strategic difference is made clear by some of the technology choices of their space program. For instance, investing in ceramic 3D printing in microgravity environments. [25] This technology is crucial for the rapid construction of structures in space using space resources, referred to as *in situ* manufacturing. This technology is tied explicitly to its intentions with the Moon, where it will "manufacture ceramic molds with the lunar dust, and then cast components using metals in the lunar soil."[26] As will be discussed later, instead of sending building materials to space, this technology only requires sending the manufacturing tools; the Moon provides all the building material required.

The ability to manufacture *in situ* using space resources will allow for exponential growth of manufacturing capability. This represents a threat to any power restricted to terrestrial resources and manufacturing. Even more so when wielded by a rising power like China.

Third, China benefits in this area because of its form of government. The Chinese Communist Party, or CCP, has continuity in the one-party system that runs the country. There are also significant detriments to this method of government, like groupthink and the suppression of debate and the free exchange of ideas. However, what it does allow is long-term planning and consistent direction to the national organizations. This centralized planning allows them to project their plans, funding, and research out three decades while the US is limited to roughly five years due to the government acquisition cycle. Assuming China's strategy is well-thought-out and realistically phased and resourced, this should result in a significantly more focused application of national resources.

Fourth, China's strategy extends beyond simply the number of researchers and amount of money invested, though that is significant. What is more concerning is the direction of their technological development. The direction and purpose are important. Putting all those resources into less worthwhile paths would not be threatening - like the US Army maintaining horse cavalry units long after they had been proven obsolete by armor and mechanized forces. They only disbanded them in 1942.[27] During World War II, horses proved to be virtually worthless when compared to the steel steeds that the armored troops wielded. Earlier investment in technology and tactics for armor would have likely increased the US's armor capabilities in the war. The most concerning technologies China is investing in are SBSP, 3D printing, artificial intelligence (AI), and quantum computing.[28]

SBSP, previously mentioned, is the crux of China's strategic plan, and is required for domestic energy needs and extending its hegemonic reach. China has been spending a significant amount of time and effort attempting to improve its terrestrial solar panel industry, possibly in preparation for conversion into producing space-based solar panels. They are currently the largest manufacturer of solar panel technology, well-positioned to convert that industry into providing panels for satellites.[29] Because they have nurtured their industrial base in this area, it is prepared to drive both their military and economic power forward.[30]

3D printing radically increases the efficiency of the space industry and allows for complete *in situ* production of future space systems. There will be no need to send resources from Earth if the space industry can provide the resources, and 3D printers can convert them efficiently into end products in space. While not a capability, it significantly increases the efficiency of China's efforts in space to achieve its SBSP vision. While the space mining and industrial base develops, raw materials could be sent from Earth to fuel these 3D printers. While most rockets used today launch delicate humans or sensitive satellite transmitters or sensors, that level of precision is not necessary when sending iron or other raw material.[31] Therefore, more gross means of sending materials to space could be contemplated, supplying the raw materials until space mining develops to the point that it can supply in-space construction needs.

AI is the ability for machines to learn from both the environment and human interactions. It will be crucial for future space *exploration* and *expansion*. As probes get further from Earth, they take longer to transmit and receive due to the incredible distances involved. While some depictions of the future space industry have humans piloting construction ships and freighters, these jobs would most likely be done by learning AI. It is also concerning in the non-space arena, where AI can churn through data more rapidly than the human mind could hope to match. This allows for "big data" analysis and can aid in things like business trends and military intelligence.

Quantum computing is the future of cutting-edge coding and communication. Google claims to have a quantum computer as of October 2019 that operates 100 million times faster than any ordinary computer.[32] Quantum computers are particularly suited to generating and breaking codes, theoretically rendering current encryption defunct while generating its own invincible digital security. While fragile and requiring extensive cooling, these devices can calculate at phenomenal speed.

    If combined, AI and quantum could create an extra-human ability to react and learn from the environment. Application to space is evident, as sending probes or even full ships equipped with quantum 'brains' and advanced AI could enable rapid *exploration, expansion,* and *exploitation* of resources throughout the Solar System. For military engagements, armed ships capable of examining one trillion possible assessments per second would be able to *exclude* adversaries very effectively. Putting all the technologies together into a vignette, SBSP can provide the power to established bases of 3D printers producing products from *in situ* resources. These would be managed through quantum AI to ensure protection, efficiency, and maintenance of the facility. SBSP provides power, and 3D printers produce more SBSP satellites – a never-ending cycle of growth and *expansion,* all managed independent of human oversight.

    However, it is not China's space technology alone that comprises their strategy of space.

China has truly embraced a Whole of Government Approach, and that includes information operations here on earth and lawfare to assist in the anticipated space occupation phase. This book uses 'lawfare' in the context of the 1999 paper by two officers of China's People's Liberation Army, where they refer to it as, "a nation's use of legalized international institutions to achieve strategic ends."[33] Lawfare uses the tools of peace and turns them to indirect war, converting plows into sharpened pens. These are shaping operations meant to make future space control easier through influence and shaping the international system to be supportive of China's goals.

Information warfare is one of the tools used by China to assist in building its dominance in space. The Confucius Institute is a Chinese organization that reports directly to the Chinese Ministry of Education and spreads Chinese propaganda to competitor nation's campuses.[34] Their propaganda extends even to the primary level of education, as the organization sends mentors to underfunded public school systems and provides instruction in Chinese, fulfilling the language program requirements. As part of the training, Chinese folk stories are told, including one referencing Chinese ownership of the Moon.[35] Part of a culture is the stories it tells, down to fairy tales and nursery rhymes. China is exercising soft power to further the claims made by their space program, that they own the Moon. While Chinese fairy tales will not suddenly brainwash a generation, having that underlying claim embedded in the American cultural fabric makes it available for future Chinese exploitation. This indoctrination of the youngest generation of Americans is just one element of the Chinese program to advance their national goals of space control.

China also uses lawfare to assist in securing its future space holdings by setting the precedent for space with terrestrial conflicts like Antarctica. The frozen tundra of Antarctica and the Moon are linked by remarkably similar treaties. The Outer Space Treaty (OST) restricts any claim of national sovereignty over the Moon and forbids militarization, just like the Antarctic Treaty does for Antarctica.[36] They are also linked by similar resource situations; both the Moon and Antarctica possess rich stores of valuable assets.[37] It is for these reasons that one can be seen as precedent for the other. China is already moving to secure the key resource locations in Antarctica, establishing at least four bases in strategic areas.[38] Since the international community has been able to do little to stop China's *expansion* into Antarctica and its occupation of key locations, they will be able to do less when it is on a different celestial body entirely. If no power can hold China in check in the frozen South Pole, the Moon will almost certainly follow the same behavior of "first come, first served."³⁹

China has crafted a strategy that plays to several of their strengths and currently has a good track record of meeting their declared schedules – including meeting their deadlines for crewed space flight, putting up an orbiting space station, and sending a probe to the far side of the Moon.⁴⁰ Of particular interest is the lunar probe landing on the far side of the Moon, because that was one of the first times that China moved beyond the footsteps of the US's accomplishments and began to make their own.

China has strong motivation from one economic and one cultural underpinning, has resources and technology aligned with their strategy, and are using all tools of national power to achieve their goal. It sees more in space than satellites and support to Chinese warfighters, there are resources and economic power to be gained. However, because space industry expansion will result in improving production capability and access, if China chooses to weaponize space later, it could do so rapidly. They are focused on *expansion* into the space domain to claim ownership and influence, as well as the economic benefits that will almost certainly come from it. In short, its strategy is explicit, and it has an aggressive schedule. The time for disbelief has passed; it is time to take China seriously in space.

[1] Senate, *China in Space: A Strategic Competition?*, 90.
[2] Chen Na, "Capture an Asteroid, Bring it Back to Earth?", Chinese Academy of Sciences, June 24, 2018. Accessed 22 May 2020: http://english.cas.cn/newsroom/news/201807/t20180724_195396.shtml.
[3] Deyana Goh, "China's Space Station to be Operational by 2022", Spacetech, March 15, 2018. Accessed 22 May 2020: http://www.spacetechasia.com/chinas-space-station-to-be-operational-by-2022/.
[4] Yamei, "China unveils Follow-Up Lunar Exploration Missions", *Xinhua*, January 14, 2019. Accessed 22 May 2020: http://www.xinhuanet.com/english/2019-01/14/c_137743015.htm.
[5] Gao Ji, et.al., "Solar Power Satellite Research in China", *Online Journal of Space Communications*, Winter 2010. Accessed 22 May 2020: https://spacejournal.ohio.edu/issue16/ji.html.
[6] Chen Na, "Capture an Asteroid?". Accessed 22 May 2020.
[7] "China Aims to Explore Polar Regions of Moon by 2030", *China Daily*, September 25, 2018. Accessed 22 May 2020: http://www.chinadaily.com.cn/a/201809/25/WS5ba9f615a310c4cc775e801f.html
[8] Gao Ji, et.al., "Solar Power", Accessed 22 May 2020: https://spacejournal.ohio.edu/issue16/ji.html.
[9] Gao Ji, et.al., Accessed 22 May 2020.
[10] Lifang, "China Focus: Flowers on the Moon? China's Chang'e 4 to Launch Lunar Spring", *Xinhuanet*, April 12, 2018. Accessed 25 May 2020: http://www.xinhuanet.com/english/2018-04/12/c_137106440.htm
[11] Xiang Bo, "China to Achieve "major breakthrough" in Nuclear-Powered Space Shuttle Around 2040: Report", Xinhuanet, November 16, 2017. Accessed 22 May 2020: http://www.xinhuanet.com/english/2017-11/16/c_136757737.htm.
[12] Senate, *China in Space*, 90.
[13] Jean-Pierre Lehmann, "October 2049: The 100th Anniversary of the People's Republic of China: What will

China Look Like", The Globalist, 29 August 2019. Accessed 25 May 2020: https://www.theglobalist.com/china-economy-demographics-politics/.
[14] Gao Ji, et.al., Accessed 22 May 2020.
[15] Gao Ji, Hou Xinbin, and Wang Li, "Solar Power Satellites Research in China" Online Journal of Space Communication #16, Winter 2010. Accessed 26 May 2020: https://spacejournal.ohio.edu/issue16/ji.html.
[16] Gao Ji, Hou Xinbin, and Wang Li, "Solar Power Satellites"
[17] Peter Garretson, "Asymmetries in Projected Space Development" Unpublished Paper, 11-12.
[18] Garretson, "Asymmetries", 14.
[19] Peter Garretson, "USAF Strategic Development of a Domain" Over the Horizon Journal, 10 July 2017, 7. Accessed 26 May 2020: https://othjournal.com/2017/07/10/strategic-domain-development/
[20] Namrata Goswami, "Explaining China's space ambitions and goals through the lens of strategic culture," The Space Review, 18 May 2020. Accessed 17 June 2020: https://www.thespacereview.com/article/3944/1.
[21] Goswami, "Explaining China's space ambitions."
[22] Erin Blakemore, "The Brutal History of Japan's 'Comfort Women'," History Channel, 21 July 2019. Accessed 17 June 2020: https://www.history.com/news/comfort-women-japan-military-brothels-korea.
[23] Jun J. Nohara, "Sea power as a dominant paradigm: the rise of China's new strategic identity," Journal of Contemporary East Asia Studies, Volume 6, 2017 – Issue 2, 210-232. Accessed 17 June 2020: https://www.tandfonline.com/doi/full/10.1080/24761028.2017.1391623.
[24] Garretson, "Asymmetries", 16-20.
[25] ZX, "China Focus: China Pioneers Ceramic 3D Printing in Micro-Gravity", Xinhua, June 19, 2018 at http://www.xinhuanet.com/english/2018-06/19/c_137265536.htm
[26] ZX, "China 3D Printing.

[27] Dick Hakes, "The day the Army unsaddled its last horse", Iowa City Press-Citizen, 21 March 2016. Accessed 6 May 2020: https://www.press-citizen.com/story/entertainment/go-iowa-city/2016/03/21/united-states-cavalry-army-unsaddled-its-last-horse/82098652/.

[28] Ray Kwong, "China is Winning the Solar Space Race" Foreign Policy, 16 June 2019. Accessed 26 May 2020: https://foreignpolicy.com/2019/06/16/china-is-winning-the-solar-space-race/; Karen Chiu, "China makes its first 3D-printed objects in space" China Tech City, 7 May 2020. Accessed 26 May 2020: https://www.abacusnews.com/china-tech-city/china-makes-its-first-3d-printed-objects-space/article/3083280; Will Knight, "China may overtake the US with the best AI research in just two years," MIT Technology Review, 13 March 2019. Accessed 6 May 2020: https://www.technologyreview.com/2019/03/13/136642/china-may-overtake-the-us-with-the-best-ai-research-in-just-two-years/; Jeanne Whalen, "The quantum revolution is coming, and Chinese scientists are at the forefront." The Washington Post, 18 August 2019;

[29] Chris Baraniuk, "How China's giant solar farms are transforming world energy" BBC, 4 September 2018. Accessed 26 May 2020: https://www.bbc.com/future/article/20180822-why-china-is-transforming-the-worlds-solar-energy.

[30] Garretson, "USAF Strategic Development", 7.

[31] Garretson, "Asymmetries", 7.

[32] Bernard Marr, "15 Things Everyone Should Know About Quantum Computing," Forbes, 10 October 2017. Accessed 6 May 2020: https://www.forbes.com/sites/bernardmarr/2017/10/10/15-things-everyone-should-know-about-quantum-computing/#27ff5b171f73.

[33] Qiao Liang and Wang Xiangsui, "Unrestricted Warfare: China's Master Plan to Destroy America," Newsmax.com, 22 August 2002.

[34] Sahlins, Marshall, "China U." *The Nation.* Published 29 October 2013. Retrieved 11 April 2014.

[35] Melia Pfannenstiel (Professor at Air Command and Staff College), interview by author, 12 February 2020.
[36] "Antarctic Treaty," United Nations, 1 December 1959. Accessed 15 June 2020: https://2009-2017.state.gov/t/avc/trty/193967.htm.
[37] L.M Foster and Namrata Goswami, "What China's Antarctic Behavior Tells Us About the Future of Space," The Diplomat, 11 January 2019. Accessed 15 June 2020: https://thediplomat.com/2019/01/what-chinas-antarctic-behavior-tells-us-about-the-future-of-space/.
[38] Foster and Goswami, "China's Antarctic Behavior."
[39] Foster and Goswami.
[40] Senate, *China in Space,* 91-94.

Chapter 7
American Space Strategy

This is a protracted conflict in the form of a struggle for technological dominance. It is primarily a contest to secure relative advantage in the domain and to get to key celestial lines of navigation first and with the most capabilities. Doing so requires specific doctrine, training, technology, investments, deployments, and missions.

--Lt Col Peter Garretson

In contrast to the nearly thirty-year plan that China has presented, the US space strategy is frustratingly short-term. All the stated milestones are within the decade and, while several are promising, there has been no overarching declaration linking them together. The US is reacting to China and must develop a new strategy to force China on the strategic defensive.

A summary of the American spacepower strategy is in the National Security Space Strategy (NSSS), of which a twenty-one-page unclassified summary is available to the public.[1] There is a revision in progress, which aims to be entirely unclassified, but it has not yet been released. The

NSSS' stated objectives are, "[s]trengthen safety, stability, and security in space; maintain and enhance the strategic national security advantages afforded to the United States by space; and energize the space industrial base that supports U.S. national security."[2] The tone of these objectives are mostly strategically defensive – the focus is on safety, stability, and security, and maintaining current advantages. There are some active words like strengthen and enhance, but most of the statement appears focused on maintaining current advantages and improving on them, not doing something new. This is the mission statement of the current hegemon that does not desire to risk losing. In contrast to this, China is embarking on a new strategy that goes beyond what the US strategy is configured to accomplish. The US seeks to improve its position in orbit, China seeks power beyond the rim of Earth's gravity well.

Another document that provides a look at US space strategy is the USSF Strategic Overview from February 2019. It specifies that the Space Force will be trained and equipped for "[p]rotecting the Nation's interests in space."[3] This document represents perhaps the most holistic approach taken to spacepower and was the cause of some optimism among those believing in space development beyond Earth's orbits, often called Blue Water Space Theorists, at the time. Subsequent US Space Force

decisions have caused much of that optimism to evaporate, and Congress has requested recognized spacepower advocates to come and make recommendations to get it back on track.[4] The recommendation provided to Congress had three major thrusts:
1. A new mindset – the USSF needs to understand space-age warfare and not merely do industrial-age warfare better.
2. A new risk culture – the current culture of aversion stifles innovation and achievement.
3. A new organization – old, safe methods of doing self-serving business cannot be indulged any more. Instead, new, vibrant organizations are needed.[5]

It remains to be seen, as the first members are being added to the Space Force as of the writing of this book, what direction it will take, and whether the Space Force will take these recommendations seriously. A new theory will drive a new strategy in any domain, and space is no different. While not perfect, hopefully SDT will accelerate the process.

In contrast to the Chinese space strategy, the American has an undercurrent of parochial satellite-centric thinking that prioritizes security, stability, and maintaining current advantages. This is at least partially because the USAF space program, birthed in the contentious Cold War, almost immediately militarized. NASA, while non-military, sought *exploration*. The problem was that there was no government organization to support space *expansion*. It is this stark lack of *expansion* after the Moon landing's *exploration* that stands as a missed strategic opportunity. This has extended to the present as the US still does not embrace *expansion* in the domain, though recent NASA missions hint at a changing understanding. The US space milestones identified are:

- May 2020: X-37B with Microwave Power Beam launched (SBSP)[6]
- 2020s: Orbiting Lunar Station proposed (NASA)[7]
- 2020s: Air Force Research Laboratory is working on SBSP[8]
- 2022: VIPER lander samples water at Lunar south pole[9]
- 2024: Artemis program returns humans to Moon.[10]
- 2025: Human mission to Mars[11]

Current American spacepower strategy explicitly contains none of these significant milestones, which is likely driven by at least two causal factors. First, the democratic system of the US sees policy and goals shift with changes in the administration. There is little chance that any US agency would be able to make claims on what it will be doing in seven years due to budgetary uncertainty and the general unlikelihood of sustained political support. Second, the governmental and bureaucratic environment constrains the American space enterprise. The US military has been dominant in space for so long and had its mind shaped by the vision of space as support to air and ground forces, that it is nearly impossible to break this organizational momentum short of a catastrophic failure or direct civilian intervention.

The situation is reminiscent of the Army Air Corps and Army debating the role of aircraft in the twentieth century. The Army viewed the new domain through the lens of what it could provide to the warfighters on the front lines. The Air Force, meanwhile, saw the immense opportunities offered by the new domain and the systems designed to exploit it. Likewise, the new warfighting domain, space, demands a new mindset and methods to use it to its fullest potential. Like the Air Force's strategic focus, a vision of spacepower must animate the new USSF and give it direction beyond merely continuing the vision of the Air Force Space Command. Satellites and flags have been the space goal before, but that will not beat Chinese resources and bases. If there is a space race between the two powers of China and America, it is not a lack of technology that will cause America to lose – it will be America's lack of strategy.

Now that the respective theories and strategies have been compared, the next chapter will outline the significant expected phases of space development. After laying out the stages, the scenario will project a possible future based on the respective country's strategy.[12]

---

[1] Secretary of Defense, *National Space Security Strategy*, Unclassified Summary January 2011.

[2] Secretary of Defense and Director of National Intelligence, *National Security Space Strategy*, January 2011, 4.

[3] Congress, "United States Space Force Strategic Overview", Washington D.C. February 2019. Accessed 21 May 2020: https://media.defense.gov/2019/Mar/01/2002095012/-1/-1/1/UNITED-STATES-SPACE-FORCE-STRATEGIC-OVERVIEW.PDF, 4.

[4] Steven Kwast, "Placing the Space Force on the Right Footing" Unpublished Briefing to Congress, 26 February 2020.

[5] Kwast, "Space Force on the Right Footing"

[6] Brett Tingley, "X-37B Space Plane's Microwave Power Beam Experiment Is A Way Bigger Deal Than It Seems," The Warzone, 8 May 2020. Accessed 16 May 2020: https://www.thedrive.com/the-war-zone/33339/x-37b-space-planes-microwave-power-beam-experiment-is-a-way-bigger-deal-than-it-seems.

[7] Erin Mahoney, "Q&A: NASA's New Spaceship" *NASA.gov*, 13 November 2018 at https://www.nasa.gov/feature/questions-nasas-new-spaceship. (Accessed 15 March 2020).

[8] Kyle Mizokami, "The Air Force Wants to Beam Solar Power from Space Back to Earth" Popular Mechanics, 4 November 2019. Accessed 25 May 2020: https://www.popularmechanics.com/military/research/a29687992/air-force-solar-power-space/.

[9] Sarah Loff, "New VIPER Lunar Rover to Map Water Ice on the Moon," NASA.gov, 25 October 2019. Accessed 16 June 2020: https://www.nasa.gov/feature/new-viper-lunar-rover-to-map-water-ice-on-the-moon.

[10] Brian Dunbar, "Humanity's Return to the Moon," NASA.gov, 19 May 2020. Accessed 16 June 2020: https://www.nasa.gov/specials/artemis/
[11] Sarah Fecht, "Elon Musk Wants to Put Humans on Mars by 2025", *Popular Science,* 2 June 2016 at: https://www.popsci.com/elon-musk-wants-to-put-humans-on-mars-by-2025/. (Accessed 15 March 2020).
[12] For further reading, see *The Future of Space 2060 and Implications for U.S. Strategy: Report on the Space Futures Workshop*, Air Force Space Command, 5 September 2019.

Chapter 8
The New Space Race

Space is the Navy for the 21st-century economy, a networked economy that will dominate any linear terrestrial economy in the four engines of growth and dominance that change world power: transportation, information, energy, and manufacturing. [...] Whoever gets to the new market sets the values for that market. And we could either have the market with the values of our Constitution [...] or we could have the values we see manifest in China."

--Lt Gen Steven Kwast

The future is dark if the US and its allies do not embrace a theory that envisions space correctly and seeks to actively *expand* and *exploit* the vast resources present. China has already declared its intentions and failing to challenge its growth will put the US in an untenable position by 2036. The US must embrace Moon *expansion*, even if it raises risk to military satellites in the short-term. Failing to do so guarantees defeat.

After examining space theory and strategy from both China and the US, as well as getting an introduction to Space Development Theory (SDT), it is finally time to put them to the test in a scenario. As Clausewitz would say, a 'war on paper.' [1] The following is a step-by-step examination of the strategies currently employed by China and the US in space. The assumption is that no allies interfere in the war, and the global avoidance of another world war continues. Finally, for the purposes of this scenario, current US and China space timelines are met successfully. The simulation follows the phases outlined in Figure 7. Phase 0 is the current moment in 2020 and simply sets the stage for the succeeding phases, broken up into five- or ten-year segments for ease of analysis. In this first simulation, the black stars on the timeline designate where the US decisively loses astronautic or spacepower advantage, respectively identified with an "A" or "S."

Phase 0 – Current Situation

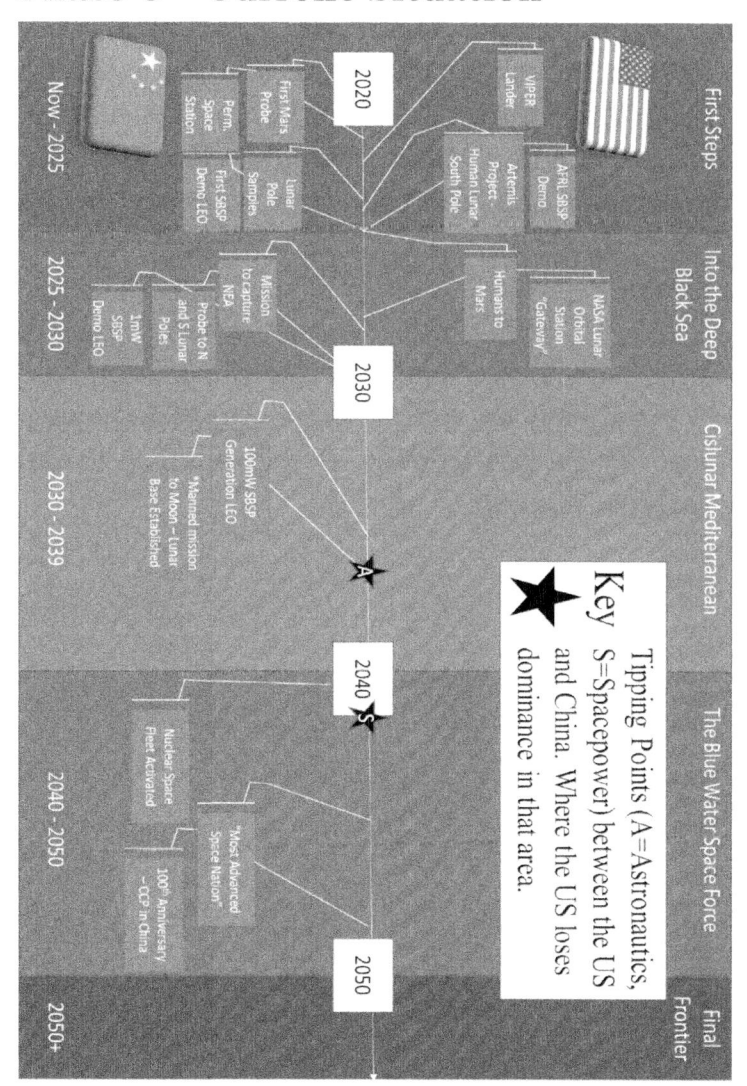

Figure 7 - The New Space Race, Current Situation Timeline Overview (original by author)

The current situation is one of American superiority in terrestrially focused space assets, with American satellites outnumbering Chinese

2-to-1. That superiority is eroding as aging American satellites reach the end of life and Chinese launches outpace American.[2]

China continues to follow its stated goals and sends the first Mars probe and finish in-orbit construction and wireless transmission for its future Space-Based Solar Power (SBSP) satellites.[3]

The USSF has begun attempting to assemble itself and embrace its purpose as the service focused on maintaining American space superiority in Earth orbit. It will achieve initial operational capability in 2022.

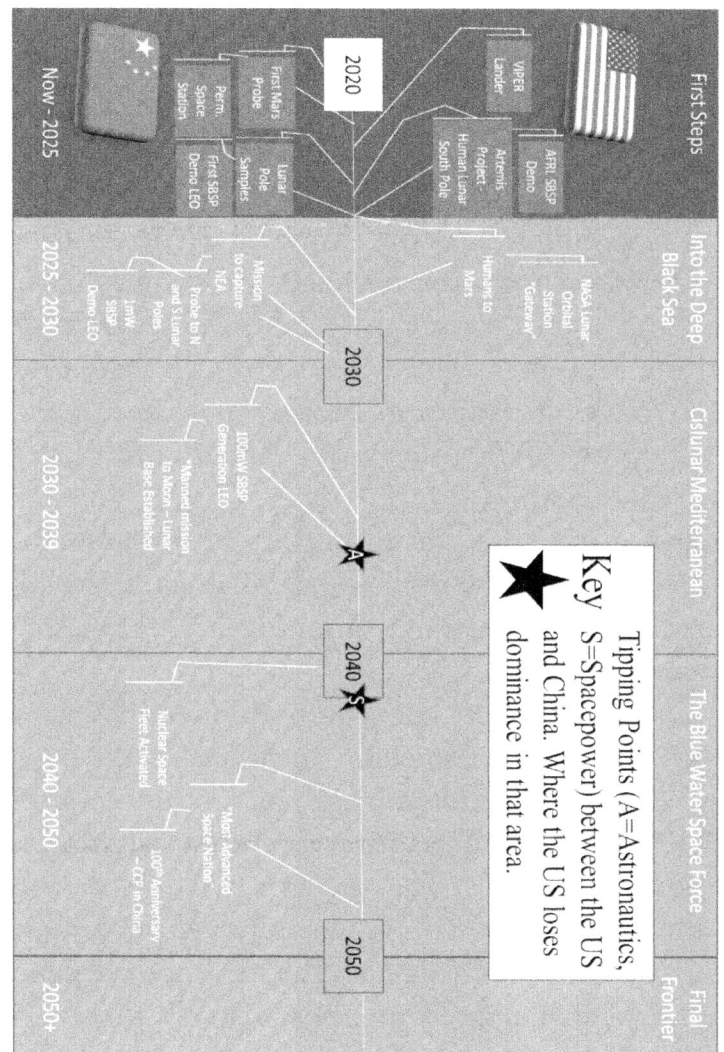

Phase 1 – First Steps (2020-2025)

Figure 8 - Phase One of the New Space Race, First Steps (original by author)

As outlined in Figure 8, China is busy, sending its first exploration probe to Mars, opening that planet for further developmental options. It also sends out an asteroid probe to begin evaluating for future asteroid-capture missions. Finally, it founds its first permanent space station, *Tianhe-1*.[4] In SDT, the Chinese are *exploring* the Moon and Mars, while they are *expanding* into terrestrial orbit with their space infrastructure.

The Americans are working on their own crewed Mars mission and begun putting a lunar base into orbit around the Moon. They continue working on Space-Based Solar Power (SBSP) and likely have a working prototype. They have *explored* cislunar and lunar space and have begun *exploring* Mars. They have further *expanded* into the various orbits around Earth. While slow, they are making initial steps to *exploit* them for societal benefit, pushing beyond purely military ends.

At this point, the competition still favors the US. Sending a mission to Mars would be a significant accomplishment for science and exploration – not to mention a first for humanity. China, in contrast, still appears to be repeating US achievements. In 2024, however, the first truly unique element of the Chinese space program will begin to manifest – the Chang'e 6 will bring back samples from the Moon's south pole.

That may not seem overtly threatening. It is some dirt from a celestial body, and the US has brought back Moon samples already and will likely have a probe there in 2022 under the

Artemis program. [5] What is concerning about China's action is not what, but where and why. The south pole of the Moon, according to Paul Spudis, is likely rich in ice.[6] Ice can be melted down into water and used to support human habitation as well as rocket servicing. [7] It is concerning because this is beginning to complete the first phase of space development for the Moon, *exploration*. Once *exploration* is complete, the next step is to *expand*. The Moon is the steppingstone that allows off-world transit; whoever holds the Moon holds the keys to the rest of the solar system.

Phase 2 - Into the Deep Black Sea (2025-2030)

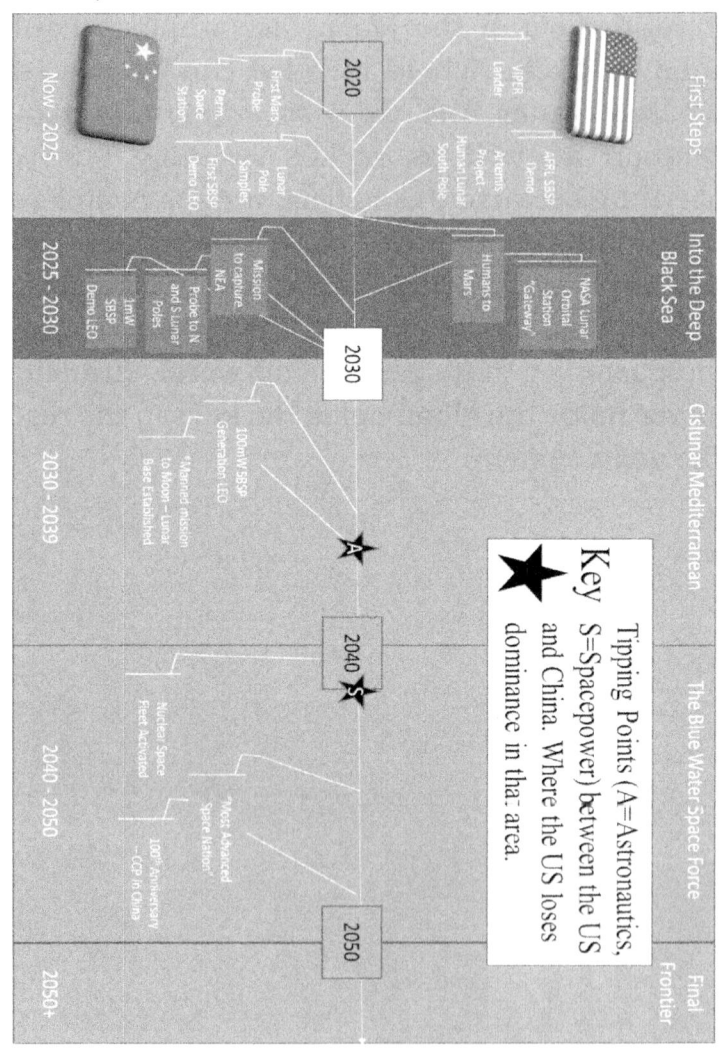

Figure 9 - Phase Two of the New Space Race, Into the Deep Black Sea (original by author)

As seen in Figure 9, Chinese *exploration* continues in several stellar "seas." In the area

100

between the Earth and Moon, called cislunar space, China will send a follow-up probe to both the North and South Lunar Poles by 2030, likely to complete a survey for a follow-on crewed station by 2036. Their Mars mission should return by 2028, which will not likely beat the US to Mars but will probably be the second one. This accomplishment will help fuel the narrative of closing the gap between the two space nations. China is also anticipating sending a probe to Jupiter in 2029, possibly *exploring* some of the moons for habitable locations. Finally, they are sending a mission in 2029 to capture a Near-Earth Asteroid (NEA). The estimated value of these asteroids is in the trillions of US dollars from iron, nickel, cobalt, and gold.[8] If the Chinese can catch some of these asteroids and return them to earth, causing them to either enter orbit or crash in an easily recoverable location, they could augment their GDP through mining.[9] They would also be able to fuel their industry with the minerals, reducing dependence on foreign sources.

    The US will continue technological innovation with its satellites and advanced technologies like SBSP. The US also will likely beat China in sending a crewed mission to Mars. It has already beaten China to Jupiter with several spacecraft, including Pioneer 10 in 1972, Pioneer 11 in 1973, Voyagers 1 and 2 in 1977, Galileo in 1989, New Horizons in 2006, and Juno in 2011. Of note, however, all mentioned spacecraft were conducting *exploration*, the first phase of domain development. In contrast, none have approached the second, *expansion*. While the US will likely proclaim its continued superiority, with good cause, this is the phase that will begin to distinguish the two strategies from each other. While the US will continue to *explore* and bolster its conventional satellite fleet around Earth; China will start seeding its industrial space capability to truly *expand* beyond Earth orbits and *exploit* the resources and locations found there.

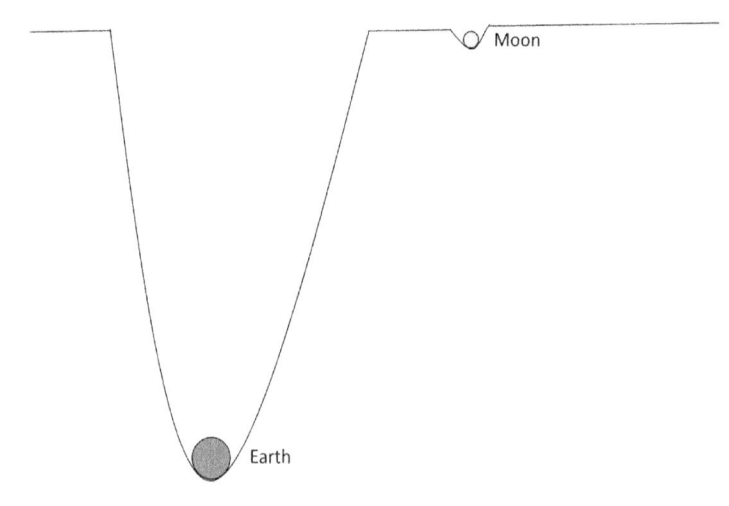

Figure 10 - Relative gravity well for Earth and Moon (original by author)

Space development can change the calculus of industry and economy on Earth forever. The most significant limiting factor in doing anything in space is Earth's gravity well. Put in terms of geography, Earth sits at the bottom of an incredibly deep valley, and the space around it is the mesa. The Moon sits in a minor ditch, by comparison. The scale of this gravity well is significant because the Earth's well is 22 times deeper than the Moon's. Figure 10 graphically illustrates this. If the plan is to stay in the valley, then the depth is mainly irrelevant. However, if the idea is to move goods on the mesa that surrounds it, it is 22 times harder to get from Earth to the mesa – that is, open space - than from the Moon.

Rocket propellant mass is currently a major limiting factor in space launch. If space launch moved to the Moon, every ounce of propellant becomes 22 times more efficient due to the lessened relative gravity.[10] If space production allowed the creation of rockets in space, that is on the 'mesa' itself, it is possible to be even more efficient. Simply put, it is incredibly propellant-expensive to transfer weight from Earth to space, but to move mass from space to Earth is easier and significantly cheaper in propellant. Looking ahead to the next phase, when the Chinese space industry begins to draw its points together in a web, the danger is that the fastest and most efficient industry on Earth will simply be unable to compete with a moderately efficient one based in space.[11]

Phase 3 - Cislunar Mediterranean (2030-2039)

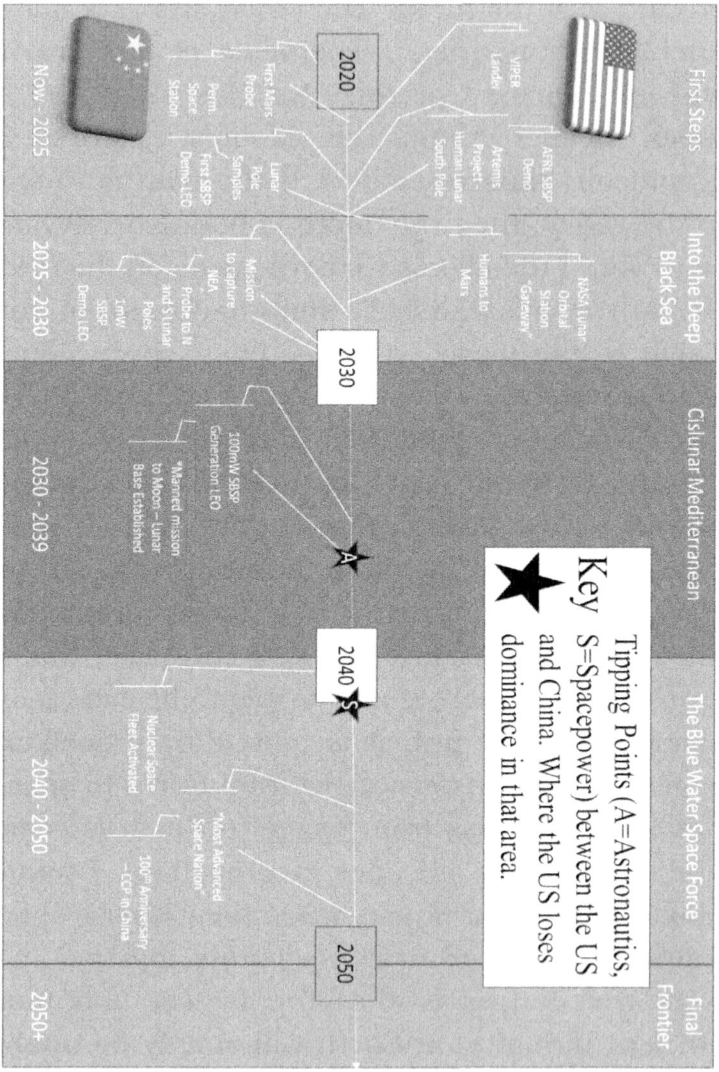

Figure 11 - Phase Three of the New Space Race, Cislunar Mediterranean (original by author)

As shown in Figure 11, it is in this phase that the Chinese strategy bears decisive fruit in astronautics. In 2034, China returns a NEA to Earth, a lunar research base *expands* China's influence through 3D printing technology, and SBSP is generating 100 MW from orbit. These elements are significant because industry relies on six factors; 1) raw resources, 2) a location to work on them, 3) power, 4) tools, 5) skill to convert them, and then 6) the ability to move the finished goods to the end destination. The Chinese will have all of these in place by 2039. They will have the ability to gather raw resources in the form of asteroids and lunar material. They will have a location to work on them, either in small orbital factories or at their fledgling Moon base. They will have the power, through their SBSP satellite, to provide a constant flow of energy from the sun. They will have the tools and skills through either 3D printing and artificial intelligence (AI) or some other solution they develop. Finally, they can move them to different destinations through the rockets that will already be in space. Even a moderately-sized nation would be able to compete against the US space production with a 22 times handicap, and the Chinese will likely be the leading global economy at this point in history.[12] The US will have a minimal ability to interfere with this new industry in space through sanctions or other traditional means. The Chinese have moved firmly from *expansion* into *exploitation* in the cislunar space. They are dangerously close to assuring their dominance into the foreseeable future, able to prevent others from following them.

Perhaps the most dramatic unipolar moment in the past hundred years occurred in 1989 with the fall of the Berlin Wall and the collapse of the Soviet bloc as a viable counterbalance to the US and NATO. The subsequent period lasted roughly twenty years until China was able to recover from losing 45 million to starvation under Mao, and Russia had time to recover and take stock of the US abilities and develop countermeasures. [13] For approximately 20 years, the US was the sole world superpower but, with the return of great power competition, that is no longer the case.

The cause of the unipolar moment was the collapse of the opposing bloc and did not require war. Even without the USSR's internal collapse, the computer revolution would have likely exacerbated the already present disparity between the Soviet bloc's industrial military and the US's computer-driven military. The upcoming space revolution will be different from the steam engine, automobile, or computer revolutions. Unlike those previous revolutions, it is the space revolution's combination of technology and location that makes it so transformative. The strategy that China has embarked on – a combination of advanced space technologies, AI, space-based solar power, space industry, and *expansion* – will change the world like the previous technological revolutions. Driving the revolution would be an advantage to any nation. For a would-be great power that is seeking global hegemony, it will likely prove decisive.

The US will likely continue to persist in *exploration* operations during this time, establishing a lunar base like the Chinese, but using it for *exploration* and not *expansion* or *exploitation*. The reason these moves will probably be insufficient is not due to the relative hardware or technology, but rather the lack of strategy that drives it. The US is responding to a nation that has planned these moves for more than ten years.

Phase 4 - The Blue Water Space Force (2040-2050)

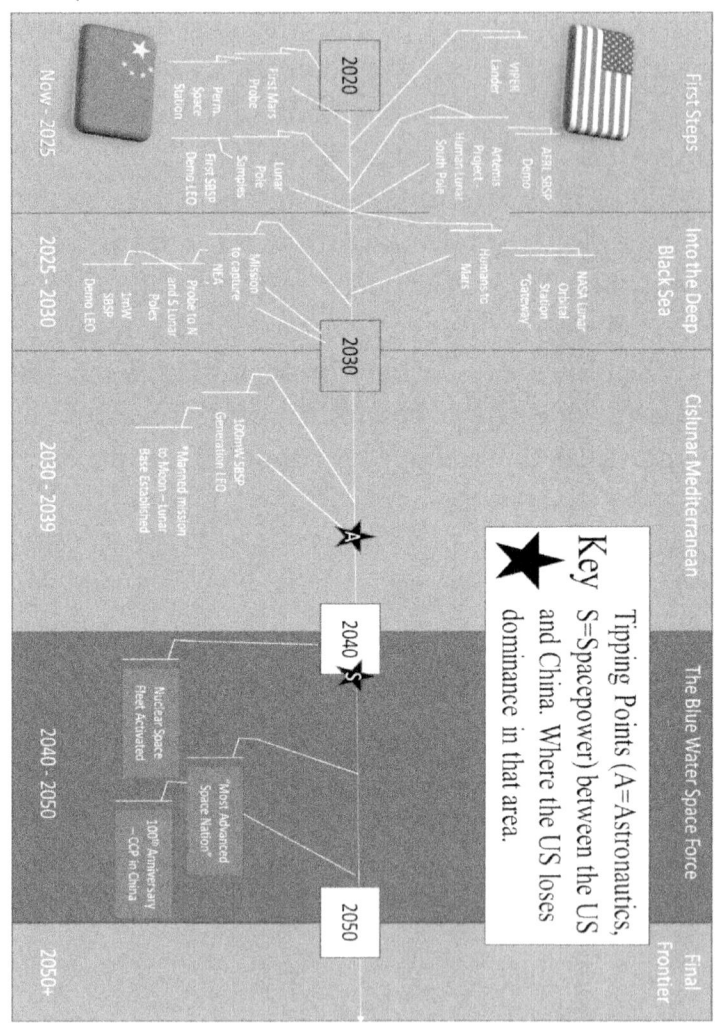

Figure 12 - Phase Four of the New Space Race, The Blue Water Space Force (original by author)

In this phase, China cements its decisive edge and surpasses the USSF in spacepower for the first time. This is depicted in Figure 12. In

2040, the most dramatic event in this timeline will occur – the christening of a Chinese nuclear-propelled space fleet.[14] In the same year, the Chinese SBSP will begin full service in GEO orbit, likely providing power to China's national users and possibly in-space production facilities. By 2049, the 100th anniversary of the establishment of the PRC, China declares itself as the most advanced space nation. In 2050, it has its first commercial SBSP satellite that can provide power, at a price, anywhere in the world.

With the maturation of SBSP, China can use the energy produced to fuel its civil and space infrastructure. According to some projections, this could be on the order of 450 GW per year, adding up to $225B to China's annual revenues if sold.[15] China can increase its SBSP to the point that it is entirely energy-independent within six years. Continuing that trend, it can export excess energy production capacity, reducing the possible leverage of other powers threatened by their expansion.[16] SBSP is the lynchpin of China's astronautics strategy, and superior strategic positioning on the Moon enables it.

In addition to SBSP bolstering China's astronautics, the space fleet formed to defend it will likely prove the most dramatic and impactful from a national security perspective. While it is impossible to predict what form it will take precisely, it could be similar to the Orion program that the USAF began in the 1960s.[17] At the height of nuclear competition between the US and the USSR, the USAF was developing a spacecraft that used atomic detonations for propulsion. This "nuclear pulse propulsion" was theorized to enable a significantly larger and more combat-capable craft to escape Earth's gravity and serve as a combat vessel in space.[18] The original plan was for more than 100, ranging from nuclear bombers in LEO to *4,000-ton* space battleships able to engage in multi-year missions to Saturn.[19] The theory was that, in the event of a nuclear war, these ships would serve as the final guarantors of Mutually Assured Destruction – raining their bombs upon the offending party.[20]

From a military perspective, nuclear pulse, or the much cleaner nuclear thermal propulsion, offers the ultimate in second-strike capability for nuclear deterrence. For more everyday uses, it also frees the space arm from Earth-based threats in all but the most exceptional situations. Modern satellites operate predictably within range of ASATs, lasers and directed energy weapons, as well as many abilities that attack in the electromagnetic spectrum. Put simply, satellites that occupy the "ultimate high ground" are finding themselves held at risk from the space equivalent of surface-to-air missiles (SAMs). If the satellites stay in orbit around the Earth, like canoes circling an island, they will remain vulnerable to the firepower directed from that island. However, once the space navy becomes indeed "blue water" and can maneuver in deep space, the silent sea, the military threat from the island becomes relatively inconsequential.[21] It is also a new weapon that will nullify many of the old. ASATs, aircraft, cyber threats, and other tools in the US DoD arsenal will likely be of little help in countering the Chinese nuclear space navy. The next British empire will have its home country on Earth, but the sun will genuinely never set on this empire. There is no horizon in space.

For expansionist ends, a nuclear-powered ship would significantly increase the mass to propellant ratio, enabling actual settlement ships to locations like the Moon or Mars. Founding self-sustaining space settlements would ensure a nation's interests persist beyond even the destruction of Earth in some cataclysm, whether natural or human-made. This capability is worthy of consideration because it can enter a nation's deterrence calculus. If one nation has space settlements and one does not, the risks each are willing to accept are different.

The US, in this phase, falls behind and is unable to compete further with the Chinese designs for three reasons. First, the Chinese economy will be superior to America's with the augmentation of the space-based resources and delivery through *exploitation*. Money can buy technology and innovation, and China will be able to purchase needed capability with its resources due to its superior economic position.

Second, the Chinese will have secured an excellent position through careful *expansion*. If

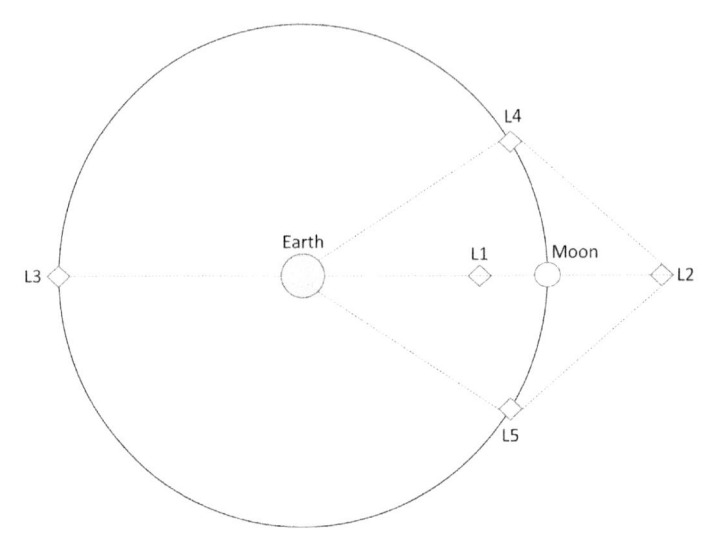

Figure 13 - Lagrange Points relative to Earth and Moon (original illustration by author)

China can lock down the Lunar Lagrange points, particularly L2, L4, and L5, and significant sections of the south Lunar ice caps first, it is possible the US may not be able to stake significant claims to any of these critical points. Lagrange Points are shown in Figure 13. Once China has secured them, it can fortify them as it has the SCS islands. At that point, the only way to acquire them is through a war, a war that the USSF would be ill-positioned to win. This preemptive occupation by China's assets secures the strategic ground, and only military force would be guaranteed to remove it. This overtly hostile act would certainly draw censure from the China-supporting international community and make the US appear the aggressor.

Third, China will be able to use the combination of these other advantages to *exclude* the US from embarking on a similar strategy because it has already *explored* and *expanded* to critical strategic locations. Once China has it, recent history shows, it will not relinquish it. Instead, it will build on its superior strategic position, fortify its gains, and look for further advantage in economic and military function.

With the Chinese space industry producing goods and services for both Earth and settlements on the Moon, their *expansion* continues unabated. At the same time, the US must still send rockets out of Earth's gravity well, incurring the same prohibitive cost to build those large enough to house sufficient propellant. The production of the first Chinese nuclear vessel would likely be a shock to America, like the USSR's launching of Sputnik in 1957. It will transform concerns into a palpable terror as the relative power imbalance comes into focus. Once the Chinese have blue water spacepower, the US will scramble to match their military might, but gravity is working for the Chinese by this point. The tyranny of the gravity well remains.

The US will likely produce some spaceships at this time, as it must attempt to maintain parity. Still, they will probably either be smaller or of insufficient quantity to assure equivalence – to say nothing of superiority – when compared to China's fleet. To put it in terms of SDT, China has reached the phase of *exclusion* in cislunar space. They have an asymmetric capability that they can leverage to use to coerce others to engage in actions they desire. This superiority is particularly damaging, as cislunar space is the Straits of Gibraltar into the silent sea. If closed, there is no other way to space and, therefore, no way to compete with the Chinese there. China has secured its dominance for the foreseeable future.[22]

SpaceX is projecting to fly to Mars around 2022 and then every two years after that to begin a Mars settlement.[23] By 2040, they could have a functioning settlement that is relatively self-sustaining. With the advent of China's Nuclear Fleet, SpaceX accepts China's protection of its Martian settlement from hostile satellites and other competitors, both commercial and national. With no American ability to protect its commercial interests abroad through *expansion*, China co-opts American commercial fruits.

Phase 5 - The Final Frontier (2050+)

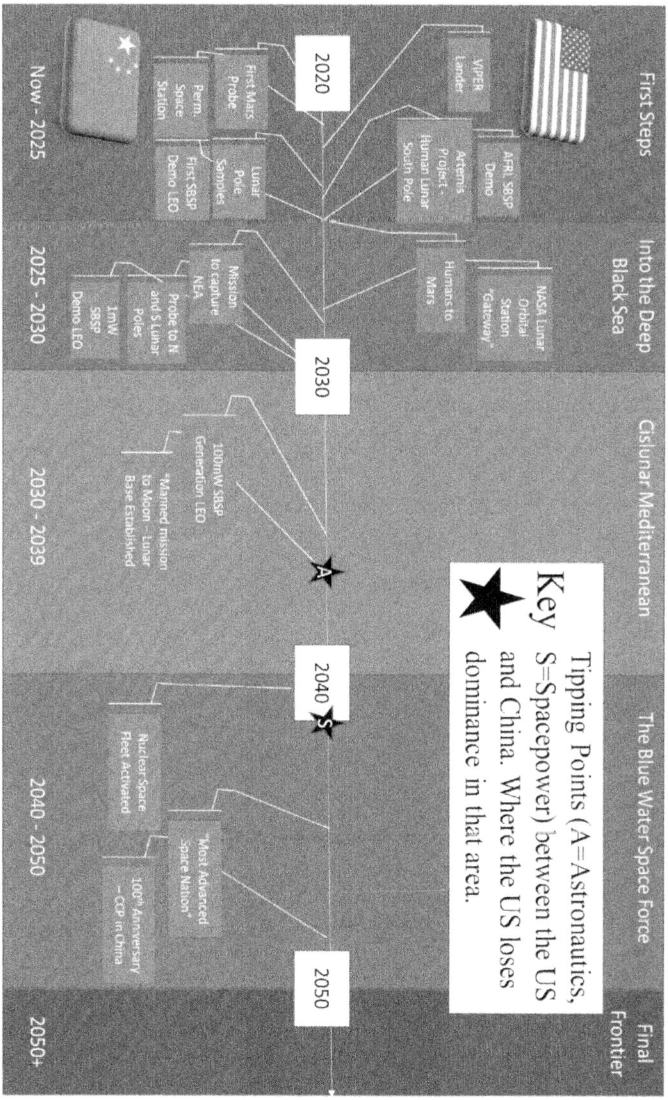

Figure 14 - Phase Five of the New Space Race, The Final Frontier (original by author)

Beyond the horizon of 2050, there is further *exploitation* possible and, if no national power can

115

challenge China in space, China will likely continue to grow unchecked. While there are no milestones beyond 2050 identified in congressional testimony, and therefore none in Figure 14, there are several resource nodes *exploitable* by China, that include the Moon, Mars, and the asteroid belt and beyond, to Uranus.

The ice present at the Lunar Poles is the most valuable resource garnered from the Moon, indeed "[w]ater is the most useful material in space."[24] Humans occupying a Moonbase would need food, hydration, and hygiene capabilities – all of which water is crucial in providing. Without unrealistically efficient recyclers or some other miracle technology, there will be no long-term occupation of the Moon unless Moon water can be gathered, refined, and used for that purpose. It would be theoretically possible to bring the tremendous amount of water required from the Earth but only at exorbitant expense. Costs to put a pound of payload into orbit ranges between $43,180 and $27,000.[25] This cost is prohibitive as the recommendation for humans is to drink 100 fluid ounces per day, weighing approximately 6.5lbs.[26] A crew of ten would be drinking 65lbs of water, equal to between $1.8M-$2.8M *per day* without recyclers. Gathering water and refining it on the Moon is far more efficient and a necessary step in Moon habitation. Any Moon habitation would need some lead-time to have robotic rovers gather and prepare water stores before the astronauts arrived.

In addition to the occupancy requirements satisfied by water, breaking it apart into oxygen and hydrogen converts it into rocket propellant.[27] This Moon *in situ* propellant production will likely be one of the first Moon industries to emerge and should be fully formed by 2050. Because there are billions of tons of water frozen on the lunar poles, the scope of this new industry will be massive and propel the rest of humanity's expansion into the Solar System.[28] Further, the propellant produced on the Moon will be more efficient due to location, as leaving the Moon's gravity requires significantly less energy than leaving Earth. Producing a rocket that can escape Earth's gravity and make it successfully to the Moon, if there is a servicing capability on the Moon, would then enable that rocket to go anywhere in the Solar System with a virtually full propellant tank. The opportunity to accelerate *exploration* and *expansion* through this increase in efficiency is difficult to overstate.

Finally, the Moon's poles receive nearly constant sunlight, which has the dual advantages of constant temperature and constant energy. The poles have relatively stable temperatures, averaging -50°C, that would enable structures to be built and maintained with relative ease when compared with the equator, which swings between -150°C and 100°C.[29] To power a polar habitation, solar power would be available around the clock because of constant exposure. It is also possible to augment this with SBSP. The solar power would provide for lighting inside the base, heating, energy for propellant production, charging for rovers and other lunar vehicles, and communications. While it is likely to have a backup generator because backups are always useful in space, the base could likely operate exclusively on solar energy. Assembled, the Moonbase provides habitation, rocket propellant, constant power, and many other benefits to *expand* beyond the cislunar area and out to the next step in *expansion*, the Martian surface.

After *expansion* to and *exploitation* of the Moon is accomplished, Mars will likely be the next destination. Unlike some advocates suggest, Mars should not be first the *expansion* as there is no genuine opportunity to *expand* there until the Moon supports significantly more efficient spaceflight. Spudis made a clinical observation of current Mars missions when he said,

> The dirty little secret is that most politicians love human Mars missions not because they have any desire or interest in *doing* them but because it is an excellent and proven way to keep the space community pacified by selecting a goal that is so far into the future that no one will be

held accountable for its continuing non-achievement. What a remarkable accomplishment for America's efforts in space: once we had a real space program that some thought was faked, and now we have a fake space program that many believe is real.[30]

Mars does have its place but as another steppingstone for further *expansion* beyond Martian orbit. Between landing and leaving, there are two critical tasks to be accomplished on Martian soil. The first is to establish a base like the Moon and accomplish the production of one of the essential apparatus for further *expansion*, the Martian Skyhook. A skyhook is a structure that extends from the surface of a celestial body into space. Climbing the skyhook, an object at its tip reaches escape velocity by moving far enough from the celestial body to escape gravity's grasp – inertial force on steroids. [31] Figure 15 is an artist's rendition of a skyhook on Earth.

It is possible to set up a base on

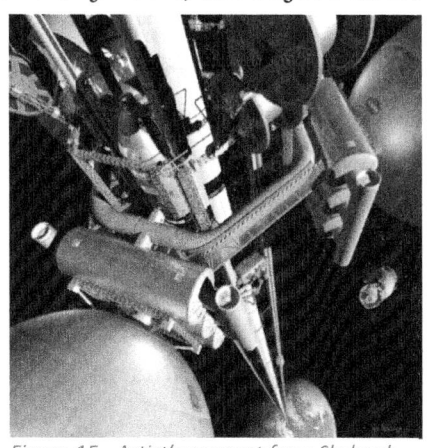

Figure 15 - Artist's concept for a Skyhook (credit NASA archives)

Mars that could generate propellant for rockets with *in situ* resources. This base would be needed, as the requirement to leave Mars to return to Earth is several times higher than leaving the Moon.[32] Carbon dioxide, one of the most common resources on the Martian surface, can be converted through high-temperature electrolysis into oxygen and carbon monoxide.[33] Using that mixture creates electrical energy and the chemical product provides propellant for rockets. Of course, a Martian base needs power to generate these propellants, and that would require either nuclear or solar power.[34] Since water is not present on the Martian surface in any known amounts, human habitation of the Martian base is unlikely until water is captured from Phobos, or hydrogen is brought from Earth, in which case this lightest element could be combined with the oxygen present on Mars and generate water on the surface.[35]

More important than the Martian base is the skyhook constructed by harnessing the two moons of Mars, Phobos and Deimos. These are remarkable for two reasons: first, Phobos is estimated to contain trillions of gallons of water – enough for servicing millions of rockets and, second, both moons are in Martian Stationary Orbit (MSO), moving at the speed of the red planet's rotation.[36]

Water has already been recognized as the most valuable resource in the Solar System when it comes to spaceflight and space development, as it serves essential functions for both human survival and locomotion in space. Phobos has nearly unlimited water reserves available if a base can extract and separate the water into propellant components.[37] This combination of skyhook and propellant would allow a rocket flying from Earth to airbrake in the Martian atmosphere and take on a full propellant load before making a trip anywhere else in the Solar System. No messy trips to Mars' surface and incurring the penalty of leaving Mars' gravity well.

Of at least equal importance to the propellant, is the possibility of building a skyhook from the Martian surface out to either Deimos or Phobos, with Phobos being the most efficient due to the colocation with water resources. To imagine a skyhook, imagine one string dropped from Phobos down to the surface of Mars, and another string extended from Phobos out into space. With a winch or some other simple mechanical device on this rope, someone could simply ascend on the rope until it left the orbit of Phobos and escaped Martian gravity entirely, sailing off into space without ever having to fire a thruster.

Of note, the original proposition was to build this on Earth. However, the gravity on Earth is stronger than Mars and would require building cables stronger than anything currently producible in bulk.[38] A Moon skyhook is practical and Liftport, a US company, is currently seeking funding to create the first one.[39] Such a space elevator would enable shipments from the Moon to be sent back to space via mechanical means instead of chemical propellant in rockets. This would help significantly with further *exploitation* of Lunar resources to reduce propellant requirements. What is lacking in the Lunar space elevator is inertial forces to throw the rockets since the Moon does not rotate. This lack of rotation means that the Moon is not an ideal solution for the *expansion* skyhook that Mars can host, helping spread humanity to the outer planets.[40] Only Mars, in the inner Solar System, is suitable for the construction of a skyhook able to send ships out to the asteroid field and beyond.

The configuration of the skyhook would allow supplies to move from a base on the surface to the propellant plant on Phobos and then up to the receiving or servicing tip at the limits of Martian gravity, allowing an entire supply chain to move propellants up to spaceships coming by to prepare for their next trip – out to the asteroid belt. If the desire were to send material back to Earth, that would only require waiting for the hook, reaching the proper location, and at precisely the right moment, pushing off. The ship would then sail to Earth with virtually no propellant expenditure.

A further advantage of Phobos is the opportunity it provides to produce space solar power at reduced costs. Phobos has the water, as already mentioned, to produce propellant for satellites and rockets. In addition to that, Phobos also has the metals required to produce solar panels with iron and nickel.[41] Mining resources in space and producing solar panels, propellant for rockets, and other materials for space based solar power satellites allows for significantly more efficient launch from Earth – possibly a mere fraction of the satellite would need to be brought up through Earth's gravity and the final product could be assembled in orbit.[42] This could continue to grow the capacity to provide electrical power anywhere in the world at around $0.09/kW-hr according to some estimates.[43] Having access to the resources on Phobos may be the next most important step, after the Moon, in space development.

After the *expansion* on Mars, the real prize of space is the resources found in the asteroid belt and outer planets, most notably Uranus. To *exploit* them in any meaningful way, however, requires the capabilities developed through Moon and Martian *expansion* and *exploitation*. Once these bases are secured, astronautics and spacepower can project out to the asteroid belt, and *exploitation* can begin there.

Asteroids that enter Earth's atmosphere and survive reentry are generally poor quality, possessing only minerals that are common on Earth, other than troilite.[44] What differentiates these asteroids from the bountiful riches found in the belt is the age and processes which formed them. The majority of asteroids in the belt appear to be 'carbonaceous chondrites' and contain significant amounts of volatiles like oxygen, hydrogen, sulfur, and others – up to 40% of the asteroid's mass.[45] If melted down, the asteroid's remaining mass will be 30% iron-nickel, which is crucial for further production and space *expansion*.[46] The asteroids in the belt have large amounts of both keys to future space development – volatiles and construction material, and they have them in such vast quantities that the human mind has difficulty comprehending it.

Estimates on resources in the asteroid belt are 825 quintillion tons of iron. For perspective on how massive this amount is, two comparisons. The first comparison is to raise the standard of living for people living on Earth. If the goal were to raise the world to US-standards of living, it would require 2 million tons of iron and steel every year.[47] The asteroid belt mining operations would provide enough to raise the entire world's standard of living for 400 million years, assuming the population remained constant and did not recycle an ounce of the iron gathered.[48] To properly *exploit* this iron for sustainable improvement of the world, mining, transportation, refining, and production operations would need to be finetuned to keep pace with annual requirements.

Another comparison is the required resources to terraform Mars, long a dream of space enthusiasts. To prevent the escape of oxygen, water, and other critical gasses, a giant dome around the entire planet could be built, turning Mars into a planetary greenhouse. Estimates are that this would take a massive amount of material, nearly a trillion tons of iron, to build this structure.[49] The belt asteroids would not even feel the loss of that much iron, as it would only be one-*millionth* of the available resources from *exploitation*.[50] The amount of iron locked in the asteroids is massive and capable of being turned to many notable projects while propelling the economies of the nations that have the foresight to *expand* and ensure they can *exploit* these resources.

Additionally, complaints about the Earth's overcrowding would be a thing of the past if the iron in the asteroid belt were gathered and construction undertaken to produce O'Neill cylinders.[51] Asteroid belt iron will provide enough material, gathered in space, and convertible without the energy penalty of traveling out of Earth's gravity well, to create space for people numbering in the quadrillions.[52] Overcrowding will likely never be an issue again for humanity with that much space and resources to maintain it.

The asteroid belt has oxygen, hydrogen (and therefore water), and iron in abundance that the human mind has difficulty grasping. Once humanity attempts to progress beyond the asteroid belt, solar power becomes too weak to be useful.[53] While solar is very practical in near-Earth space and stays viable out to the asteroid belt, it suffers diminishing returns beyond that. What is needed is a clean, readily available, and massive source of energy to propel humanity's quest past the asteroid belt to the outer planets.

What is needed is Helium-3 (He^3) – an incredibly rare element in the solar system that provides energy through exceptionally clean nuclear fusion. The Sun generates it but mining directly from Sol is a technical impossibility and the He^3 radiated from the Sun is so diffuse that it would be nearly impossible to catch enough to make gathering it worthwhile.[54] The gas giants of Jupiter and Saturn have significant amounts, but their gravity ensures any trip there with current technology – or likely plausible technology – would be decidedly one-way.[55] Uranus, however, has tons of He^3, and it is technically feasible to gather it with relative ease. Returning it to Earth would not be overly complicated, as even small amounts would be incredibly valuable. Indeed, due to the incredible energy it contains, it is one thousand times more valuable per pound than gold, valued at $16 million per kilogram.[56]

The energy contained in He^3 is massive – with the entire energy consumption of the Earth coming in at 8,500 gigawatts, which would require 450 tons of He^3 per year to run the entire world on this energy source exclusively.[57] That seems like a large amount of He^3. However, the current annual expenditure of coal alone is 3,732 million tons of oil equivalent in 2017.[58] In comparison, He^3's weight is a paltry mass to expend for the energy gained. Furthermore, the amount of He^3 present on Uranus that is capable of being gathered with current technology is more than enough for four *billion years* at current rates of expenditure.[59] With that resource harnessed, the energy future of the entire human population need never be in question again, if shared equitably.

Another significant benefit of having access to a supply of He^3 is, if possible, even more revolutionary than securing energy for humanity beyond their likely possible needs – rocket engines. Helium-3 is both clean-fusing and incredibly efficient, and some studies have already suggested it as the best option for developing new and powerful rocket engines.[60] These new engines would be able to cut the current trip to Mars from the current plod, which could take almost a year, to a sprint of as little as two months.[61] A round trip between Uranus and Earth could be cut from six years to two.[62] Additionally, because the relative mass of the propellant is small, rockets and spaceships using this propellant would be able to have much larger cargo or living areas, increasing their efficiency even further. With He³, humanity enters a new period with a revolutionary power source for both domestic and space development.

Fueled by this power source and plentiful solar power, virtually inexhaustible supplies of water, carbon, silicates, metals, oxygen, and all the space that could possibly be desired, China continues to *expand* uncontested. Like an eternal ripple, China's influence and capability continues to spread out, increasing exponentially and ensuring the world, even if allied against China, cannot hope to match its manufacturing and military capability. [63] In this dark future, an ascendant China remakes the world system in its image, imposing its "social credit system" on the rest of the world through economic and diplomatic leverage. [64] Under this system, China is able to ban people flying or taking trains, banning children from "bad credit parents" from good schools, restricting jobs, banning from hotels, and having pets confiscated. [65] With no hope to compete in space, and the massed resources on Earth unable to compete with the capabilities that space provides, the US is unable to halt the remaking of the world system around China and the values they hold.

Therefore, the space competition between the US and China is neither academic nor inconsequential. The nation, and its allies, that can *expand* and *exploit* these resources will wield power to shape the international order. From markedly improved space operations provided by the water found on the Moon and Mars, the iron in the asteroid belt, and the crown jewel of He^3 on Uranus, the *exploitation* of the resources of the Solar System will provide unparalleled power. Scrambles for resources have usually provoked violent incidents as competing nations struggle for access to propel their respective economies. This struggle occurred during both European expansions – the "New World" of the Americas and Africa in the 1800s. The new space race will likely be no different. However, China has already laid out a plan to *expand* and *exploit* the near-Earth resources before moving to the further reaches described in this section. Instead of being a domain that can be readily contested by others, space as a domain is ascendant. It opens the door to other worlds and resources that are at the limits of humanity's mind to comprehend. China has set itself up to *exploit* this new terrain, and, like the homesteaders from American history, they plan to stay. Unlike those homesteaders, there is no shining sea to contain this expansion on the final frontier.

---

[1] Carl von Clausewitz, *On War*, ed. and trans. Michael Howard and Peter Paret (Princeton, NJ: Princeton University Press, 1994), 119.
[2] Ivan Couronne, "In space, the US sees a rival in China" Phys.org, 6 January 2019 at

https://phys.org/news/2019-01-space-rival-china.html. (Accessed 15 March 2020).
[3] Senate, *China in Space: A Strategic Competition?*, 90.
[4] *Ibid.*, 90.
[5] Sarah Loff, "New VIPER Lunar Rover to Map Water Ice on the Moon," NASA.gov, 25 October 2019. Accessed 17 June 2020: https://www.nasa.gov/feature/new-viper-lunar-rover-to-map-water-ice-on-the-moon
[6] Spudis, 120-21.
[7] *Ibid.*, 123-24.
[8] Lewis, 193.
[9] Joel Sercel, "Industrializing the Near-Earth Asteroids: Speculations on Human Activities in Space in the Latter Half of the 21st-Century," presented 3-4 April 1990 at NASA Lewis Research Center, Cleveland, Ohio. Accessed 11 June 2020:
https://ntrs.nasa.gov/archive/nasa/casi.ntrs.nasa.gov/19910012828.pdf
[10] For further discussion on cheaper space launch that uses terrestrial options, see *Fast Space: Leveraging Ultra Low-Cost Space Access for 21st Century Challenges*, Air University, Maxwell AFB, AL, 22 December 2016.
[11] Secure Freedom Radio, "Lt. Gen. Steven Kwast: Space is going to be the economy of the 21st century" Center for Security Policy, 6 February 2020 at https://www.centerforsecuritypolicy.org/2020/02/06/lt-gen-steven-kwast-space-is-going-to-be-the-economy-of-the-21st-century/. (Accessed 15 March 2020).
[12] Albert Keidel, "China's Economic Rise – Fact or Fiction": Policy Brief 61, (Washington, D.C.: Carnegie Endowment for International Peace, 2008), 1.
[13] Frank Dikotter, "Mao's Great Famine: Ways of Living, Ways of Dying." Dartmouth University. Accessed 17 June 2020:
http://www.dartmouth.edu/~crossley/HIST5.03/FILES/OHMC_DIkotter.pdf
[14] Xiang Bo, "China to achieve 'major breakthrough' in nuclear-powered space shuttle around 2040: report" Xinhua.net, 16 November 2017. Accessed 22 May 2020:

http://www.xinhuanet.com/english/2017-11/16/c_136757737.htm

[15] Peter Garretson, "Asymmetries in Projected Space Development", 14.

[16] Peter Garretson, "Asymmetries," 14.

[17] George Dyson, *Project Orion: The True Story of the Atomic Spaceship* (New York, NY: Henry Holt and Company, 2002.) 214.

[18] Brent Ziarnick, "Tough Tommy's Space Force: General Thomas S. Power and the Air Force Space Program" (Maxwell Air Force Base, Alabama, Air University Press, April 2019), 78.

[19] Dyson, *Project Orion*, 181-184.

[20] Ziarnick, "Tough Tommy's Space Force", 97.

[21] Brian E. Hans, Christopher Jefferson, and Joshua Wehrle, *Movement and Maneuver in Deep Space: A Framework to Leverage Advanced Propulsion,* Air University Press, Maxwell AFB, May 2019.

[22] Dolman, *Astropolitik*.

[23] Elon Musk, *Making Life Interplanetary,* abridged transcript of speech delivered 28th September 2017 at 68th International Astronautical Congress in Adelaide, Australia, 8.

[24] Spudis, 123.

[25] Sarah Kramer and Dave Mosher, "Here's how much money it actually costs to launch stuff into space." Business Insider, 20 July 2016. Accessed 7 May 2020: https://www.businessinsider.com/spacex-rocket-cargo-price-by-weight-2016-6.

[26] Gina Shaw, "Water and Your Diet: Staying Slim and Regular with H20" WebMD, 7 July 2009. Accessed 7 May 2020: https://www.webmd.com/diet/features/water-for-weight-loss-diet#1.

[27] Spudis, 199-200.

[28] Spudis, 1.

[29] Spudis, 122.

[30] Spudis, 127, italics original.

[31] John S. Lewis, *Mining the Sky: Untold Riches from the Asteroids, Comets, and Planets.* Basic Books, New York, New York, 1997, 181.

[32] Lewis, *Mining the Sky*, 161.
[33] Lewis, 161.
[34] *Ibid*, 161-163.
[35] *Ibid*, 168
[36] *Ibid*, 179-180.
[37] *Ibid*, 179.
[38] Victor Liu, "A New Way of Traveling in Space: The Skyhook" The Spectator, 25 January 2020. Accessed 8 June 2020: https://www.stuyspec.com/science/a-new-way-of-traveling-in-space-the-skyhook
[39] "Going Back to the Moon and Our Lunar Space Elevator infrastructure", Liftport.com, 2018. Accessed 8 June 2020: http://www.liftport.com/lunar-elevator.html.
[40] *Ibid*. 181.
[41] Leonard Weinstein, "The Solution to Future Energy Needs and Global Pollution," the Air Vent, 19 April 2009. Accessed 8 June 2020: https://noconsensus.wordpress.com/2009/09/02/the-solution-to-future-energy-needs-and-global-pollution/
[42] Weinstein, "Solution to Future Energy Needs"
[43] Weinstein.
[44] Lewis, *Mining the Sky*, 191.
[45] Lewis, 193
[46] *Ibid*, 193.
[47] *Ibid*, 194.
[48] *Ibid*, 194.
[49] *Ibid*, 194.
[50] *Ibid*, 194.
[51] Gerard K. O'Neill, *The High Frontier: Human Colonies in Space.* (Apogee Books, Canada, 2000), 37-44.
[52] Lewis, *Mining the Sky*, 194.
[53] Lewis, 204.
[54] Lewis, 205.
[55] *Ibid*, 208.
[56] *Ibid*, 210.
[57] *Ibid*, 210-211.
[58] Dennis Coyne, "World Coal 2018-2050: World Energy Annual Report (Part 4)" (Peak Oil Barrel, 20 September 2018) Accessed 8 May 2020:

http://peakoilbarrel.com/world-coal-2018-2050-world-energy-annual-report-part-4/.
[59] Lewis, 211.
[60] *Ibid*, 211-212.
[61] Fraser Cain, "How Long Does it Take to Get to Mars?" (Universe Today, 9 May 2013) Accessed 8 May 2020: https://www.universetoday.com/14841/how-long-does-it-takc-to-get-to-mars/); Lewis, *Mining the Sky,* 213.
[62] Lewis, 213.
[63] Brian Wang, "Exponential industrialization of space is more important than combat lasers and hypersonic fighters," Next Big Future, 26 October 2017. Accessed 11 June 2020: https://www.nextbigfuture.com/2017/10/exponential-industrialization-of-space-is-more-important-than-combat-lasers-and-hypersonic-fighters.html.
[64] Alexandra Ma, "China has started ranking citizens with a creepy 'social credit' system – here's what you can do wrong, and the embarrassing, demeaning ways they can punish you," Business Insider, 29 October 2018. Accessed 11 June 2020: https://www.businessinsider.com/china-social-credit-system-punishments-and-rewards-explained-2018-4
[65] Ma, "China's creepy 'social credit' system"

Chapter 9
Toward an American Space Strategy

> *Control of space means control of the world. If out in space, there is the ultimate position from which total control of the Earth can be exercised, then our national goal and the goal of all free men must be to win and hold that position.*

--Senator, later President, Lyndon Banes Johnson

Lessons learned from the "war on paper" that did not go well for the US include the economic power of space, the key importance of not merely *exploring* but also *expanding*, and a re-examination of the purpose of spacepower. The decisive edge in the new space race will not be military technology, but economic production capability.

There is a gap between the Chinese space strategy and the current US one, at least as it has been communicated *writ large*. To provide clarity, consider the following assertions of astronautics and spacepower:

1. **Astronautics' decisive effects are through economic power.** Decisive here is not in the traditional military sense of decisive military engagements that bring about the destruction

of the enemy's army and subsequent surrender. Rather, space, and the control of it, will be decisive in the great power competition through economic means. Being able to deter a potential adversary successfully can still be considered decisive because the highest form of strategy is achieving goals without fighting.[1]

2. ***Expansion* must follow *exploration*.** The US has adapted well thanks to forward-thinking individuals anticipating change and setting up the organization to benefit from it. Hap Arnold is one example, the great airpower advocate from WWII. He integrated training, production, doctrine, and organizational design in a truly exceptional way. He paved the way for what the USAF became.[2] Reimagination and creative thinking is the key, not merely getting better at the current strategy. Procuring better or more defensible satellites works as a temporary band-aid to improved competitor capabilities but does nothing once the locus of conflict moves to cislunar and beyond. The US is currently locked in a reactive strategy of trying to counter competitor moves instead of changing policies. The US must have an *expansion* mindset and encourage it in the USSF, NASA, and other space organizations.

3. **Taking space's key terrain first is essential.** There are places to *expand* that provide strategic benefits just as Gibraltar and Singapore did for the British in their naval

expansion. The north and south poles of the Moon, the Lagrange Points, among others, are crucial to secure and hold against possible opponents. These all secure and improve space lines of communication (SLOC) and allow denial of the same to the enemy. These must be secured with a permanent presence before others can do the same. Preemptory occupation can be a powerful strategic tool, as any move to unseat the occupation appears overtly hostile while, in actuality, is a response to an offensive move.

4. **Space can project to Earth easier than Earth projects to Space.** This deserves some explanation. This reality is not the case today as satellites are held at risk by both ground based ASATs and co-orbital attack systems.[3] However, once there is an independent industrial base in space that can source its own material and produce manufactured goods, this reverses. At that point, holding space against Earth will become easier than attempting to hold Earth against space. Several physical phenomena benefit the exercise of power from space to earth, most notably the gravity well that allows objects to drop to Earth with minimal effort. Sending an object from Earth requires amazing technology and expenditure of propellant and other resources. Moving an object to Earth requires a fraction of the propellant and thrust to drive

or redirect it. The gravity of Earth works in space's favor, making even relatively small asteroids into *de facto* WMD. What is necessary is the *expansion* and *exploitation* of space, culminating in microgravity manufacturing and operations in the blue water of space. Once that is accomplished, space can be held against Earth.

5. **Militaries do more than "kill people and break things."** Most modern militaries have specialized in warfighting to the point that doing other things is alien to them. Many American military personnel had significant resistance to engaging in peacekeeping operations.[4] Militaries defeat enemies, but they also project power. The US Army has *explored*, built public works, and smoothed the flow of commerce in the westward *expansion* of the US. The Navy, too, has done "non-military" things like charting ocean currents, which benefitted Atlantic commerce, and providing soundings of the continental shelf that prevented commercial shipwrecks.[5] The USSF mindset must be holistic – aiding the *expansion* of the other elements of national power by establishing infrastructure to encourage movement into key areas of the domain and actively protecting assets there. Militaries serve more purposes than merely fighting wars, needlessly restricting them to that is myopic.

6. **Joint Warfighters are beneficiaries, not drivers, of space control.** Modern Joint Warfare requires space capabilities to be effective. Position, navigation, and timing (PNT), communication, intelligence, missile warning, and weather are all provided by spacepower in its current configuration. The USSF should avoid none of these. Like the USAF before it, it will give the necessary support to warfighters. However, the first goal of any domain-specific military service is to dominate that domain. Once that is complete, then the USSF can support the joint fight. Joint warfighters are beneficiaries of securing space – but securing space must always be the first goal of a Space Force in both war and peace.
7. **Individual innovation cannot overcome national failure.** As the SpaceX example from the scenario showed, hoping that technological innovation will overcome national blindness is foolish. Ultimately, businesses are motivated by profit because it helps ensure survival in the context of a competitive marketplace. Businesses are also generally very risk-averse to losing employees' lives. A global economy can draw business away, and another country that can offer protection when the home country fails to do so has a strong negotiation position with the company in question. If the US has not

expanded into the relevant space competition areas, and China can offer rescue to stranded vessels, protection from piracy, and guarantee the safety of off-world settlements, companies will ingratiate themselves with China. A company always competes with other companies. If one company suffers piracy and another does not, the immune one will likely win the competition.

All these observations point to a commercial and power-projection focus in space strategy that goes far beyond the current satellites orbiting Earth. The issue with being the *status quo* power is that the US wants to conserve what it already has. It is motivated to counter adversary moves but is not motivated to make decisive moves of its own. This positional difference appears to explain some of the variations between American and Chinese space moves over the last three decades. America continues to be the preeminent space power in technology, numbers of systems, and capability. China, on the other hand, has a tangible goal to build toward.

The typical response from the US is, "But the Chinese cannot innovate." This has been put forward by significant figures like Joe Biden in 2013, "China – and it's true – is graduating six to eight times as many scientists and engineers as we have. But I challenge you, name me one innovative project, one innovative change, one innovative product that has come out of China."[6] It is not dissimilar to other myopic self-deceptions; it delivers comfort but does not help provide either analysis or prediction.

The Chinese do innovate, along both the technological and strategic axis. They have developed both new uses for old technology, such as repurposing lasers to target air and seacraft that they desire to deter but not destroy.[7] It develops new technology as well, such as their new quantum transmission system on one of their latest satellites.[8] If they cannot develop technology, they will acquire it through theft – as evidenced by their gutting of a competing company to acquire the 5G technology to start Huawei.[9] This theft of technology was the first threat from China to the US' space industry identified in a recent Air Force Research Lab report.[10]

Most importantly, they innovate strategically by incorporating the lessons of theory and history. The US cannot afford the opiate of deception anymore, it needs to adopt a bold strategy that does more than sustain current capability. The US needs to embrace the potential of astronautics and spacepower instead of being shackled to the accomplishments of the past.

---

[1] Sun Tzu, *The Art of War*, 77.

[2] Dik A. Daso, *Hap Arnold and the Evolution of American Airpower*, Smithsonian History of Aviation Series (Washington, D.C: Smithsonian Institution Press, 2000).
[3] Theresa Hitchens, "Russia Builds New Co-Orbital Satellite: SWF, CSIS Say," Breaking Defense, 4 April 2019. Accessed 10 June 2020: https://breakingdefense.com/2019/04/russia-builds-new-co-orbital-satellite-swf-csis-say/.
[4] Laura Miller, "Do Soldiers Hate Peacekeeping? The Case of Preventive Democracy Operations in Macedonia" Sage Journals, 1 April 1997 at https://journals.sagepub.com/doi/abs/10.1177/0095327X9702300306?journalCode=afsa. (Accessed 15 March 2020).
[5] Jason Smith, *To Master the Boundless Sea* (Chapel Hill, University of North Carolina Press, 2018).
[6] Dexter Robert, "Biden Makes a Habit of Dissing Chinese Innovation" Bloomburg Businessweek, 29 May 2014 at https://www.bloomberg.com/news/articles/2014-05-29/biden-makes-a-habit-of-dissing-chinese-innovation. (Accessed 15 March 2020).
[7] Patrick Cronin and Ryan Neuhard, "China's military and paramilitary forces have been employing lasers with increasing frequency since at least 2018," The Diplomat, 2 April 2020. Accessed 17 June 2020: https://thediplomat.com/2020/04/countering-chinas-laser-offensive/.
[8] Gabriel Popkin, "China's quantum satellite achieves 'spooky action' at record distance" Science Magazine, 15 June 2017 at https://www.sciencemag.org/news/2017/06/china-s-quantum-satellite-achieves-spooky-action-record-distance. (Accessed 15 March 2020); Liu Zhen, "How China's military has zeroed in on laser technology" South China Morning Post, 4 May 2018 at https://www.scmp.com/news/china/diplomacy-defence/article/2144756/how-chinas-military-has-zeroed-laser-technology. (Accessed 15 March 2020).
[9] Steven Kwast, interview by Dr. David Livingston, *The Space Show 3461,* 14 February 2020.

https://www.thespaceshow.com/show/14-feb-2020/broadcast-3461-lt.-gen.-ret-usaf-steven-kwast

[10] Thomas Cooley, Eric Felt, and Steven Butow, "State of the Space Industrial Base: Threats, Challenges, and Actions," Air Force Research Laboratory: Defense Innovation Unit, 30 May 2019, 2.

Chapter 10
American Space Strategy Tomorrow

A vision without a strategy remains an illusion.

--Lee Bolman

After the more theoretical observations from the last chapter, it is time to propose a new strategy based on them. A theory, an understanding of the domain, is key but insufficient if it is not used to develop a strategy - the chosen method of applying resources to objectives. There are four areas that need to be addressed in the strategy: national purpose, space-based solar power (SBSP), Moon basing, and nuclear-powered spaceships.

The US needs a good strategy based on a good theory to pursue the current competition with China. It must encompass more than just the USSF, as the USSF is only one branch of the military, and incapable of winning the upcoming struggle on its own. There must be a genuinely whole-of-government focus on getting ahead of China in the *expansion* and *exploitation* of space resources. The National Space Council, reestablished in 2017 after being disbanded in 1993, could serve as an excellent coordinating authority for driving space policy across all organizations.[1] It already includes the Secretaries of Defense, Commerce, Transportation, the

Director of National Intelligence, the Administrator of NASA, and the Director of Science and Technology Policy.[2] What is needed is a clear mission, similar to JFK's Moon mission declaration: return to the Moon to set up a permanent settlement and lead the world in SBSP. Protection must be given to US space startups as well, shielding them from China's attempted buy outs through either encouraging commercial investment or legally preventing their purchase by extra-national entities. The rest of the development for space should flow naturally from there as the industry attempts to capitalize on the new *expansion* with *exploitation*.

With the newly gleaned lessons from the previous thought experiment, Figure 16 shows a proposed timeline of recommended American space strategic milestones. The strategy embraces the Blue Water school of spacepower and is focused on both promoting commercial advancement and developing a military capability to protect the new market. The milestones suggested here are mainly military, due to available space. There are no stars on this timeline as the US stays ahead of China throughout the scenario. This presentation is an ideal timeline where US technology remains superior due to the current lead in research and development.

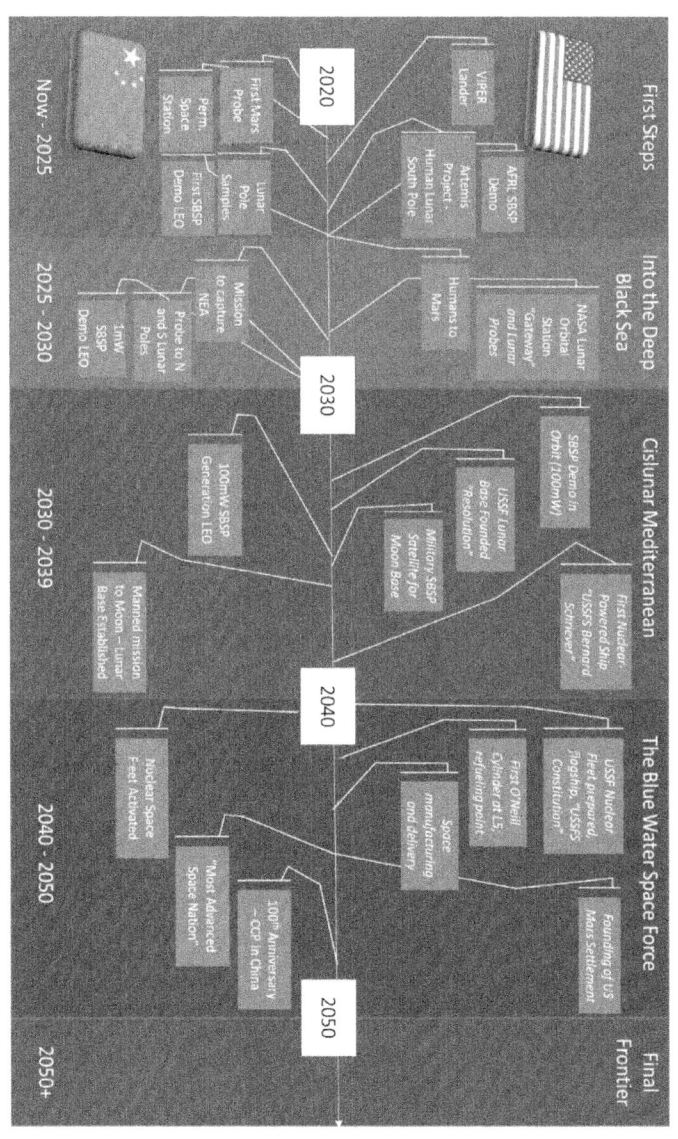

Figure 16 - The New Proposed American Space Strategy showing new milestones in italics (original by author)

Specific recommendations for the US strategy are in four main areas – national direction, SBSP, Moon basing, and nuclear-powered spaceships. The first recommendation provides a unified and focused development of the space domain to avoid the fracturing of national interests. It must be informational, cultural, and political. Informational because it must communicate to external audience what the US' goals are and that the sought control of space is for the purpose of encouraging an open market in space. Cultural because the competition will likely be generational and having development of space embedded into the society's DNA should assist in continuing support. Finally, politically, clear direction and goals need to be laid out with the dream enumerated. Fear is a great motivator but hope is a better sustainer.

The second recommendation counters China's future strategic center of gravity in space, SBSP, by competing with them in both technology and production.[3] The US is already working on SBSP prototypes, including AFRL's SSPIDR, and NRL's PRAM-FX.[4] The technology is not the issue at stake, but the purpose that drives the whole of government strategy. These technological marvels can either serve as museum pieces to US space capability or the forerunners to a new space industrial revolution. If the US is going to counter China effectively, it must be present in the domain with the appropriate tools. Since China will be using the economic power of space, the US must be present with equal or superior economic capability.

The third recommendation counters China's rapidly maturing plans to increase the efficiency of their space industry by establishing basing on the Moon and using that station for mining, servicing, and re-launch capability. This Moonbase is the astronautic tipping point for the scenario, and so allowing China sole occupancy is unacceptable. Despite the Outer Space Treaty (OST) expressly requiring all "use of outer space, including the Moon...be carried out for the benefit and in the interests of all countries, irrespective of their degree of economic or scientific development" there are serious questions of whether this prohibition will survive the first permanent occupation. [5] China is currently attempting to reshape international outer-space law, and there are indications it will emphasize, 'first come, first serve.'[6] At least five nations have made overt statements or conducted missions with obviously *expansion* goals. China, Russia, India, Japan, and the US have all made the Moon a part of their future space plans.[7]

China, as already explained at some length, has emphasized the Moon as an enabler of their future cislunar architecture plans for an expanded China.[8] The US has recently refocused NASA from a Mars emphasis to a "Moon to Mars" mission and spreading the US' presence.[9] Russia has announced plans to create a Moon base in three phases lasting from 2025-2040.[10] Japan has revectored its space program since the discovery of Lunar ice, planning on a base by 2030 and resource extraction by 2040.[11] Finally, India has stated their space program views the Moon as a "base for fuel and oxygen and other critical materials." It has already sent at least one unsuccessful probe to conduct reconnaissance at the south pole.[12]

In short, by 2040, the Moon will likely be crowded, with at least five settlements in various stages of construction around the prime real estate at the lunar south pole. These will come with zones of non-interference and other protected areas developed to avoid misunderstanding among the five anticipated space powers. The US must help craft a better legal framework for lunar settlement and resource extraction as well as build relationships now with potential allies on the Moon.

While the OST rules out "national appropriation" of celestial bodies, it does not restrict extraction and use of resources – nor does it establish who will own them once extracted.[13] The International Institute of Space Law recently stated that resource extraction is possible as long as national legislation regulates it.[14] All nations mentioned previously are at various stages of passing the required space legislation or partnering with nations that have.[15] For this reason, it is even more critical that the US provide leadership in space with a strong presence or risk being sidelined by rival powers.

India and Japan have been partners with the US over the last few decades, and their militaries exercise with the US regularly. In contrast, both Russia and China compete with the US and seek to alter the current world order. While it is not within the scope of this book to discuss alliances or coalition activities in space, there appears to be a golden opportunity here to align the India, Japan, and US programs together in opposition to the common adversaries of China first and Russia second.

In addition to the legal and alliance frameworks, to physically secure a Moonbase, *in situ* resources should be used. If brought from Earth, the materials would be prohibitively expensive in both cost and resources expended. Such a path would be needlessly inefficient. Instead, remotely operated construction should be used. This is possible because the Moon is near enough to Earth that radio signals only have a three-second delay when broadcasting between the two bodies.[16] This proximity enables radio-controlled machines to land for reconnaissance of the Lunar surface and build the first Moonbase for human habitation. Using 3D printing technology, the dirt on the Moon - called regolith - can be super-heated and fused together to create buildings. Fed through the construction equipment to build the structure, it will require thousands of thin layers of molten regolith to finish. After a system checks to ensure that the new base is airtight, it would be ready for the first astronauts to live on the Moon for extended periods. Due to the OST's restrictions on military units being on the Moon, this base would need to be decidedly astronautic in function.

Fourth, in contrast to the astronautically bend of SBSP and Moonbase, nuclear-powered spaceships have a heavy spacepower focus. Two major design branches exist. The first could see these rockets using nuclear pulse propulsion, like the Orion Project. It had a ship designed to ride the explosion of a string of hundreds of small nuclear bombs.[17] The other, significantly cleaner, approach would be nuclear thermal propulsion. This approach to nuclear propulsion has significant benefits, delivering twice the thrust with the same amount of propellant mass as traditional chemical rockets.[18] The engine superheats liquid hydrogen with a nuclear reactor. Using hydrogen as fuel allows refueling of these ships from the ice on the Moon. A sketch of how one could work is below in Figure 17.

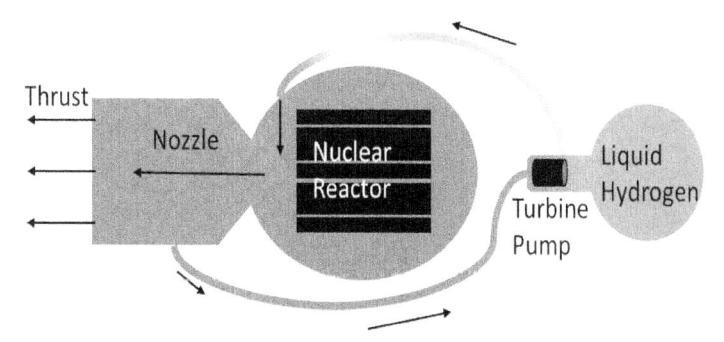

Figure 17 - Conceptual art of fission nuclear thermal rocket (original by author)

If China can produce either of these ship designs without a competitor nation, and refuel them from Moon resources, the superpower balance would likely tip in their favor around 2040. The specter of a nuclear-propelled space fleet is a significant threat to both terrestrial power and, of course, any space development that has occurred by that point. The US is already pursuing the technology and provided $100M in funding in the 2019 Federal Budget to NASA to develop space nuclear propulsion.[19] It must be pursued vigorously in the context of the strategic plan and theoretical underpinnings.

While nuclear-powered ships are dramatic and have the likelihood to shift whatever balance of power exists in 2040, it is the astronautics that will allow effective production and use. Moon basing and economic benefits from SBSP will make nuclear-powered ships more feasible and will add strategic depth to national defense plans. Without the astronautic foundation, a nation with nuclear-power ships will be in Germany's naval position in WWII. It will possess powerful warships, but with limited ability to replace losses and virtually no chance to improve national economic outlook through trade. Astronautics must be emphasized over spacepower, especially in the short term, because a navy will exist if threats to commerce are present; without that driver, the navy will wither as other interests vie for funding and personnel.[20]

Of note, this proposes strategy should be used regardless of the international situation. If China falters due to any number of internal issues, the strategy of the US in space should not change. Space has incredible, untapped potential, and another rival will challenge the US eventually – Russia has already declared its intention to set up its own Moonbase.[21] Taking and holding strategic points in space now both sets up the US for further benefit and serves as a deterrence to possible rivals.

Failing in any of the four areas identified here would be significantly detrimental to US' interests in space. China is actively challenging the US through its *expansion.* If unchecked, China will tip the balance of astronautics and spacepower solidly in their favor by 2040. Therefore, the US needs to embark on a new space race, a New Cold War, but unlike the one from the last century. Different from the 'boots and flags' that determined the winner of the

1960s race to the Moon, this new race will be for settlements, resources, and increased delivery capabilities.

---

[1] Donald J. Trump, "Presidential Executive Order on Reviving the National Space Council, 30 June 2017. Accessed 25 May 2020: https://www.whitehouse.gov/presidential-actions/presidential-executive-order-reviving-national-space-council/
[2] Trump, "Reviving the National Space Council"
[3] Office of the Chairman of the Joint Chiefs, "Joint Publication 5-0: Joint Planning" (Department of Defense, June 16, 2017), http://www.jcs.mil/Portals/36/Documents/Doctrine/pubs/jp5_0_20171606.pdf, xxii.
[4] Maj. Nathan Greiner, "Demonstration Rocket for Agile Cislunar Operations (DRACO)" DARPA. Accessed 26 May 2020: https://www.darpa.mil/program/demonstration-rocket-for-agile-cislunar-operations; Matt Jorgenson, "Project directly supports the Space Solar Power Initiative converting solar power in space to radio frequency to Earth" Northrop Grumman Newsroom, 11 December 2019. Accessed 26 May 2020: https://news.northropgrumman.com/news/features/northrop-grumman-and-us-air-force-research-laboratory-partner-to-provide-critical-advanced-technology-in-space-solar-power; Rafi Letzter, "Why a microwave-beam experiment will launch aboard the Air Force's secretive X-37B space plane", LiveScience, 15 May 2020. Accessed 26 May 2020: https://www.livescience.com/microwave-beam-military-space-plane.html.
[5] United Nations, *Treaty on Principles Governing the Activities of States in the Exploration and Use of Outer Space, including the Moon and Other Celestial Bodies*, adopted 19 December 1966, Article I.
[6] "China Expects to Introduce Space Law around 2020," China Daily, 17 November 2014,

https://www.chinadaily.com.cn/china/2014-11/17/content-18930721.htm.
[7] Namrata Goswami, "The Race to the Moon is Getting Crowded: Legal Implications," Live Encounters, June 2020, 16-17.
[8] "Exploiting Earth-Moon Space: China's Ambition after Space Station" China Daily, 3 March 2016, http://usa.chinadaily.com.cn/china/2016-03/08/content_23775957.htm,
[9] NASA, "New Space Policy Directive Calls for Human Expansion Across the Solar System," 11 December 2017. https://www.nasa.gov/press-release/new-space-policy-directive-calls-for-human-expansion-across-solar-system.
[10] "Russia Announces Plans to Establish Moon Colony by 2040," The Moscow Times, 19 November 2018. https://www.themoscowtimes.com/2018/11/29/russia-announces-plans-to-establish-moon-colony-by-2040-a63557.
[11] "The Moon-Shot Japan's Era of New Space," CNBC, n.d, https://www.cnbc.com/advertorial/the-moon-shot-japans-era-of-new-space/
[12] Elizabeth Howell, "Chandrayaan-1: India's First Mission to the Moon," Space.com, 28 March 2018, https://www.space.com/40114-chandrayaan-1.html; Department of Space, India Space Research Organisation, "Water on the Moon," https://www.isro.gov.in/water-moon.
[13] Namrata Goswami, "Race to the Moon," 17.
[14] Goswami, 18.
[15] Goswami, 19-20.
[16] Paul D. Spudis, *The Value of the Moon,* 208.
[17] George Dyson, *Project Orion,* 215.
[18] Iain Boyd, "To safely explore the solar system and beyond, spaceships need to go faster – nuclear-powered rockets may be the answer," The Conversation, 20 May 2020. Accessed 26 May 2020:
https://theconversation.com/to-safely-explore-the-solar-system-and-beyond-spaceships-need-to-go-faster-nuclear-powered-rockets-may-be-the-answer-137967.
[19] Boyd, "Safely explore the Solar System."

[20] Robert W. Shufeldt, *The Relation of the Navy to the Commerce of the United States,* Washington, DC: John L. Gink, 1878, 6.
[21] "Russia Announces Plans to Establish a Moon Colony by 2040," The Moscow Times, May 29, 2018, https://www.themoscowtimes.com/2018/11/29/russia-announces-plans-to-establish-moon-colony-by-2040-a63557.

For more information on these topics, see:
1. Paul D. Spudis, *The Value of the Moon: How to Explore, Live, and Prosper in Space Using the Moon's Resources,* 2016.
2. John S. Lewis, *Mining the Sky*, Reading, MA: Helix/Addison Wesley, 1996.
3. Gerard K. O'Neill, *The High Frontier: Human Colonies in Space,* Apogee Books and the Space Studies Institute, 2000.
4. Brent Ziarnick, *Developing National Power in Space: A Theoretical Model*, McFarland & Company, Inc. Publishers. Jefferson, North Carolina, 2015.

Chapter 11
The New Space Race, Reimagined

Having therefore no foreign establishments...the ships of war of the United States, in war, will be like land birds, unable to fly far from their own shores. To provide resting places for them, where they can coal and repair, would be one of the first duties of a government proposing to itself the development of the nation at sea.

--Alfred Thayer Mahan

Having taken the lessons from the last scenario to heart, the US embarks on a different strategy this time, embracing Moon *expansion* and economic *exploitation* of space resources to increase national power. The outcome is quite different, because the theory and strategy are different.

Like the last wargame, this one will also be a mental exercise that attempts to grasp the likely outcomes of a space race between the US and China. The independent variable is the strategy, not technology. The dependent variable is the relative industrial power of the nation-states involved. Like last time, allies are not accounted for, and a global war does not occur during this period. China's plans remain the same as in the previous scenario. The assumption is they have already embarked on the most optimal plan possible within their capabilities. In contrast, in this scenario, the US will embark on a significantly more purposefully developmental strategy that will actively challenge China's *expansion* and *exploitation* goals through focused national goal setting. More than a President declaring the vision, it is a popularly absorbed positive message. The precise and electrifying, "We choose to go to the Moon" speech by JFK, was one such moment in the past. This new strategy assumes it is captured again.

First Steps Redux (Now – 2025)

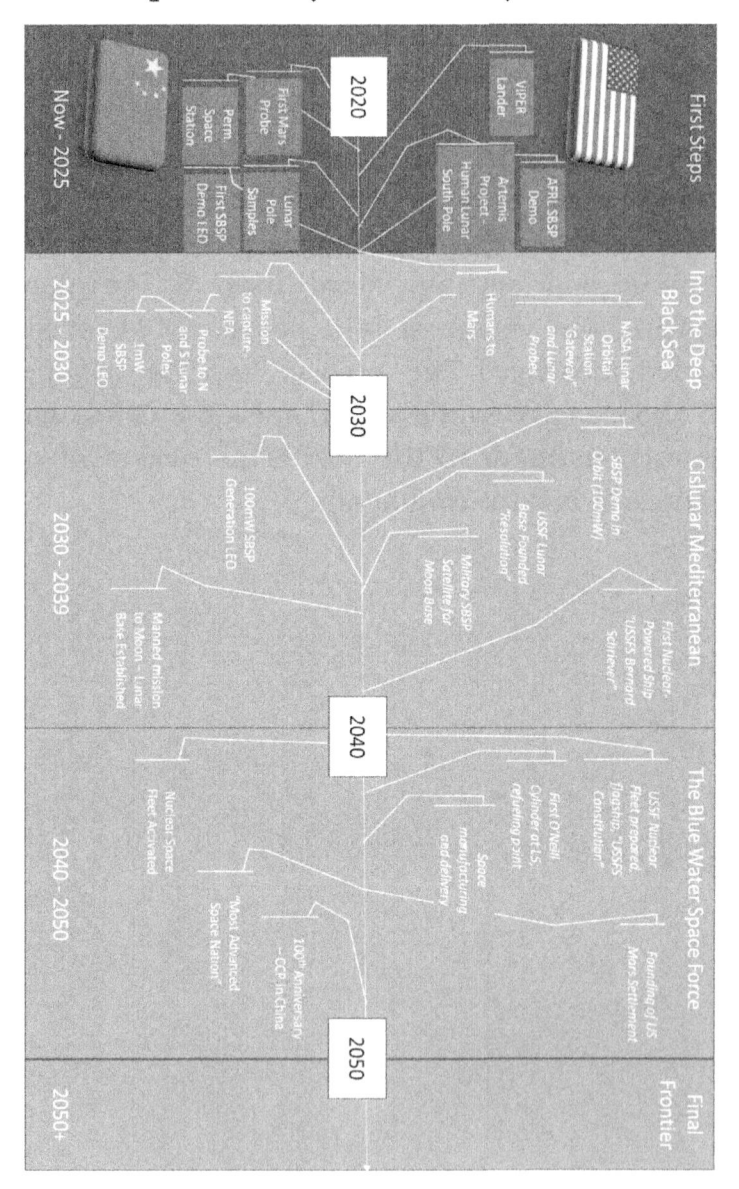

Figure 18 - The New Space Race, Reimagined, First Steps Redux (original by author)

As seen in Figure 18, China maintains its strategy and sends its first Martian exploration probe, its asteroid probe, and completes its first permanent space station, the *Tianhe-1,* as it did in the first scenario. It also *explores* the Moon and Mars while *expanding* into terrestrial orbit.

The US is still working on its own crewed Mars mission and has begun putting a lunar base, "Gateway" into orbit around the Moon. It continues working to mature SBSP technologies and has several working prototypes. It has *explored* cislunar and lunar space and has continued *exploring* Mars. It has further *expanded* into the various orbits around Earth and is beginning to *exploit* them for societal benefit. Through the Artemis Program, it also *expands* to the Lunar south pole, a significant step forward. As in the previous scenario, the competition favors the US.

Different than the last scenario, the US emplaces industrial capability at the Lunar south pole before China can establish their base. The VIPER lander, sent to the south pole in 2022, allows the US to begin *exploring* in preparation for Lunar *expansion* in the next phase. The last scenario described all the benefits derived from a Moonbase, including resource gathering and propellant efficiency for Moon launch. Mahan discussed the importance of ports for shipping, and the Moon is the ultimate spaceport to serve that purpose.[1] This move by the US is the first step changed by the new strategy in this scenario. The previous strategy is to conduct more *exploration* – the new emphasis is on *expansion* and anticipates *exploitation*.

Into the Deep Black Sea Redux (2025 – 2030)

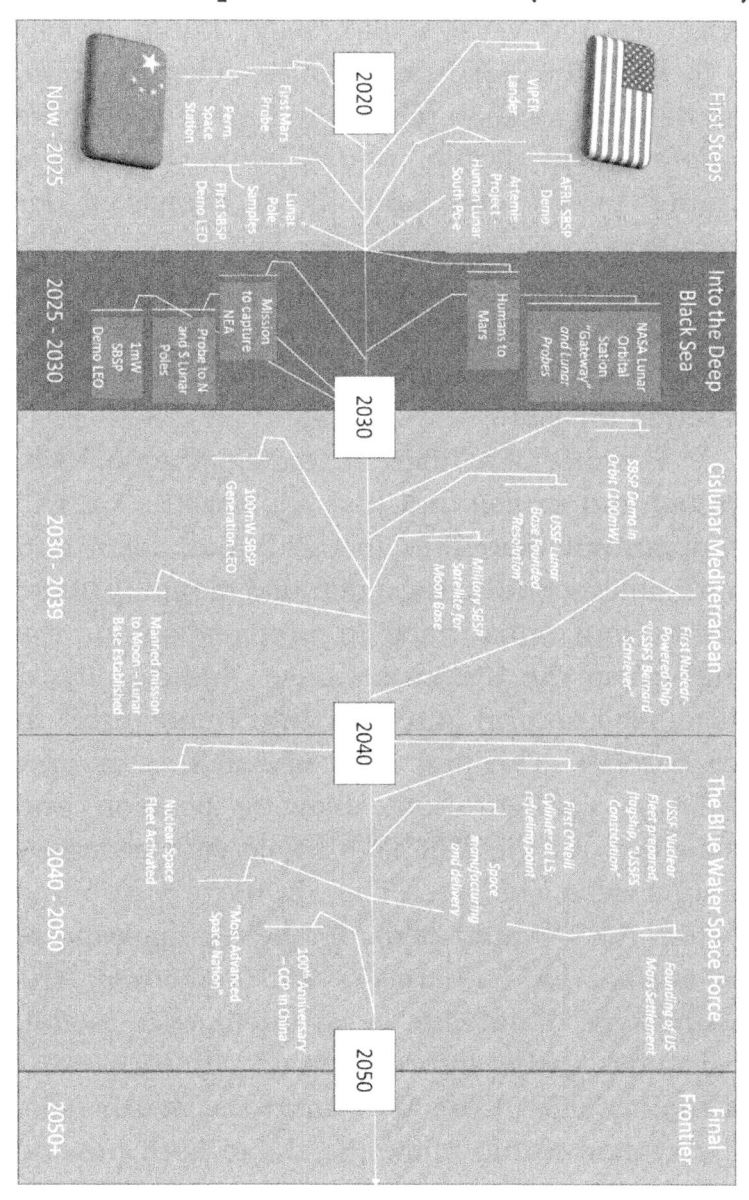

Figure 19 - The New Space Race, Reimagined, Into the Deep Black Sea Redux (original by author)

As seen in Figure 19, China's *exploration* continues in several stellar 'seas.' Lunar probes go to both the North and South poles by 2030, and its Mars mission returns to Earth in 2028. Its *exploration* in Lunar and Cislunar space is coming to an end, and it is preparing for *expansion* by 2036. Also, like last time, the US will continue technological innovation with its satellites and advanced technologies like SBSP and beat China's *exploration* to Mars by sending a crewed mission.

Because of a different strategy that looks to *expand* and *exploit*, not just *explore*, the US will send multiple satellites to the Lunar north and south poles, conducting a detailed mapping of the surface and subsurface in preparation for the founding of the US' Lunar base. Additionally, it will deploy several satellites into Lunar orbit in two batches. The first type of satellites is of a GPS-type configuration, allowing position and navigation on the Moon's surface with accuracy unachievable before. The Lunar Positioning System (LPS) would allow for better mapping of the surface for both rovers and personnel. The second type of satellite is an intelligence model, designed to gather information on China's actions on the far side of the Moon and the South Pole. This system would allow the US to both provide overwatch for its expeditions and watch China's actions to ensure it is doing as it claims. The first satellites will use direct signals from Earth for communication.

In this new scenario, the US is racing to set up architecture to *expand* to the Moon and keep ahead of China. There is no mention of China preparing a similar satellite architecture in Lunar orbit. However, the deployment of such a system by the US could provoke China to begin preparing a comparable one.

Cislunar Mediterranean Redux (2030 – 2039)

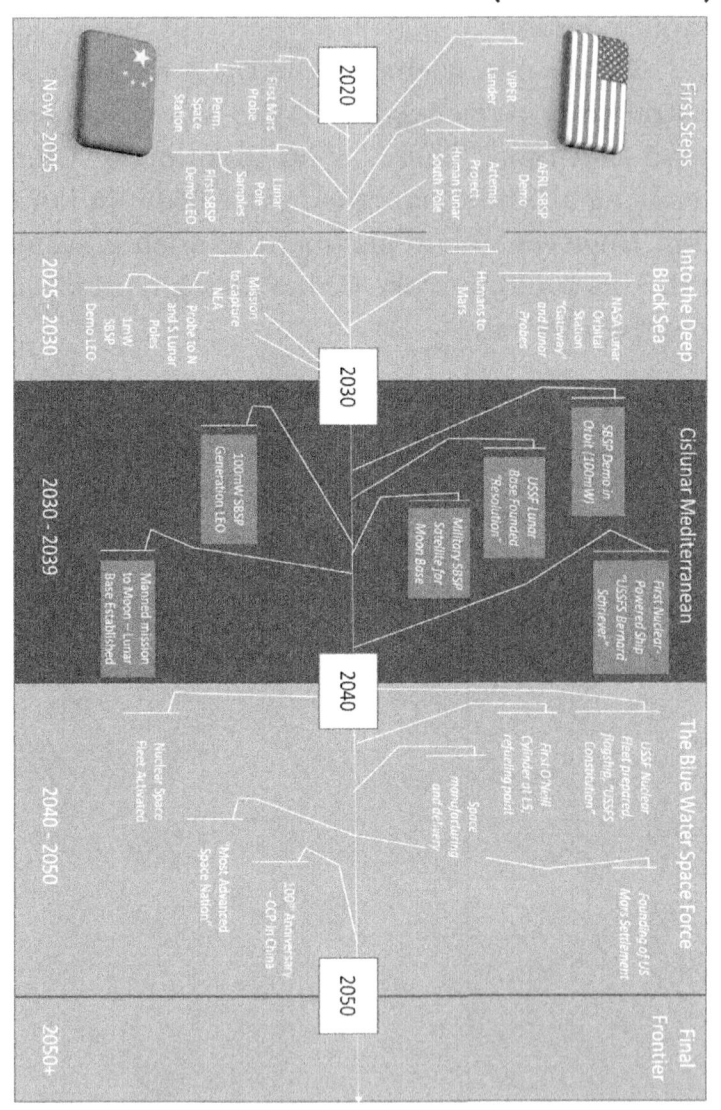

Figure 20 - The New Space Race, Reimagined, Cislunar Mediterranean Redux (original by author)

As seen in Figure 20, this was the phase that had China taking the decisive edge in the space race with their astronautics surpassing the US' with the founding of its Lunar base and the deployment of an SBSP satellite. This tipping point vanishes as the US continues to outpace China and *expands* into Cislunar space before China can successfully cement its foothold. China begins *exploitation* of Lunar resources with the founding of its Lunar Base in 2036, but now it is not the first one on the Moon's surface.

In contrast, in this scenario, the US strategy hits its stride during this phase with the deployment of its first SBSP satellite in 2033 and its own crewed Lunar base "*Resolution*" in 2034. As *Resolution* comes online, it takes command of the USSF satellite constellation orbiting the Moon, improving reaction time, and reducing the possibility of jamming from Earth-based capabilities. A second SBSP satellite in 2035, sent into Lunar proximity, likely at one of the Lagrange Points, provides sufficient energy to "Resolution" for resource gathering and launchpad construction at the Lunar South Pole. As the USSF *expands* into cislunar space, the US can *exploit* the Lunar resources under the watchful eye of USSF satellites. US commercial entities begin arriving, speeding the *exploitation* of Lunar resources, and strengthening the US' resolve in protecting its interests on the Moon. Assured protection, likely profit, and advantage against their commercial rivals - companies will almost always *expand* into new markets in these circumstances.

Unlike last time, China does not have a monopoly on the SBSP market and is unable to *exploit* it to its full potential due to US competition. While China can close its energy shortfalls with SBSP, its export market shrinks as US satellites provide similar resources at competitive prices. Because of this, China's space production goals are hampered due to less export income, forcing smaller batches of satellite and rocket production. Because the US subsidized its SBSP market, it was able to compete with China in the long-run and cause cascading impacts through both astronautic and spacepower spheres.

Finally, in a crowning achievement, the first nuclear-powered USSF space cruiser, the *USSFS Bernard Schriever,* is launched to patrol Cislunar space. Its mission is to provide security and act as a policing agent for commerce that is beginning to flow between Earth and the Moon. Hundreds of micro-satellites orbit Earth, producing further US capabilities in space from the resources gathered from the Moon, augmenting both astronautic and spacepower capabilities.

The Blue Water Space Force Redux (2040 – 2050)

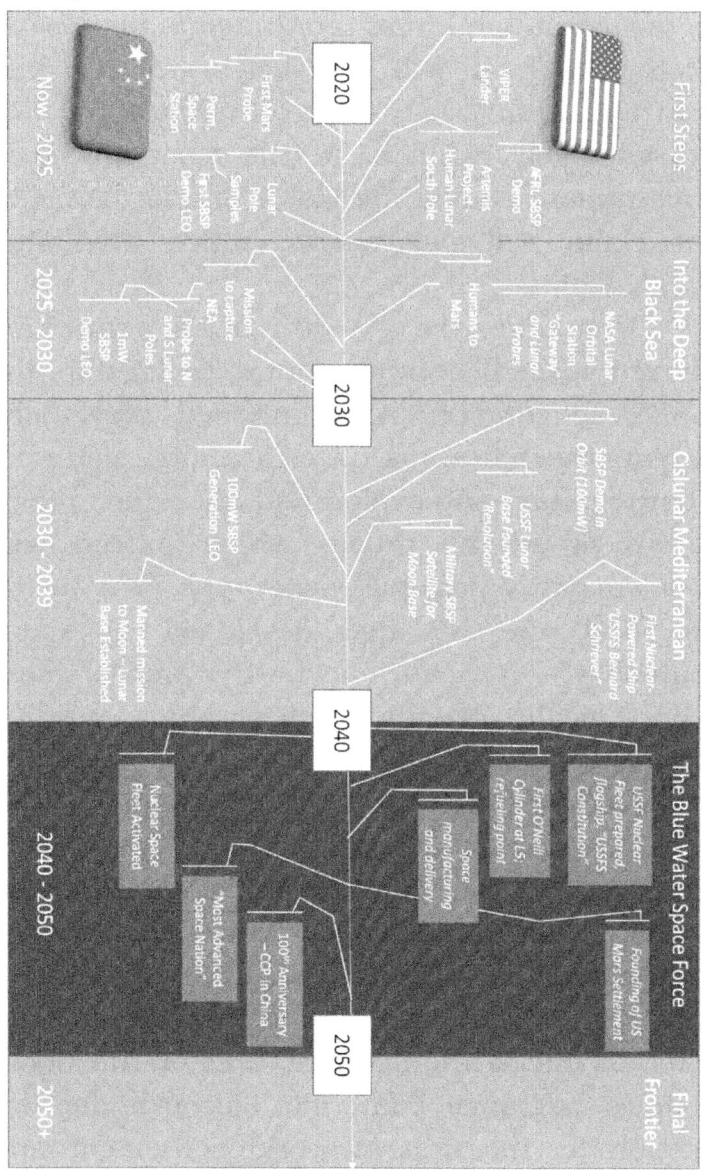

Figure 21 - The New Space Race, Reimagined, The Blue Water Space Force Redux (original by author)

As seen in Figure 21, China fields its nuclear-powered space fleet by 2040, likely being the second nation – after the United States – to do so. By 2049, it will still celebrate the one-hundredth anniversary of the founding of the PRC and will likely claim to be the most advanced space nation. Unlike the last scenario, however, that claim will not receive much credibility outside China.

This time, the US also fields a nuclear-powered space fleet built around its flagship, the *USSFS Constitution*. As mentioned previously, nuclear-powered ships have incredible potential in both military and exploration roles due to their mass-to-propellant ratios, which exceed any other currently possible rocket type. Due to the lead that the US has developed in space technology, it can field a much more technologically capable fleet of ships, holding China in check due to their superior capability and space basing

In addition to the military spacepower advantage that the US has at this point, there are several other milestones that the strategy accomplishes in this phase. First, a smaller, prototype O'Neill cylinder, constructed from materials mined and manufactured on the Moon, occupies Lagrange Point L5. This cylinder will function as the first off-world settlement and brings several benefits that include the possibility of microgravity manufacturing, off world food growth, and extremely efficient propellant use when arriving and departing.

With microgravity manufacturing, the mass of something is of significantly less importance than if it were built on Earth. Construction of much larger ships is possible without the concern of getting them out of the Earth's gravity. Food growth in space would ensure continual human presence and survival for a theoretically infinite period without ever having to return to a planet. Finally, propellant can be used more efficiently in microgravity and would allow the exchange of goods among space-based systems, whether O'Neill cylinders, space stations, or star cruisers, with minimal effort, forming the basis for a space-exclusive economy.

With this O'Neill cylinder and other in-space construction, the space manufacturing and delivery system comes into its own. Goods manufactured in space can be delivered anywhere in the world within hours, depending on the location of the actual manufacturing facilities. With 3D printing and other technologies, replacement equipment could be printed in orbit and delivered during a battle or medical emergency and brought straight to the area in need within hours.

Finally, the US inaugurates its first Mars settlement in 2049 to highlight that, while China may claim to be the most advanced space nation on its hundredth anniversary, that is not the case. With Moon launch, in-space production, and a space fleet to protect the extended lines of communications, the US can *exclude* hostile actors from areas as it sees fit. The US has won this new space race with at least one settlement on the Moon and a second one on Mars, along with an O'Neill cylinder positioned at L5. The US is prepared to move to the outer solar system and reap the vast material rewards found in the asteroid belt.

The Final Frontier Redux (2050+)

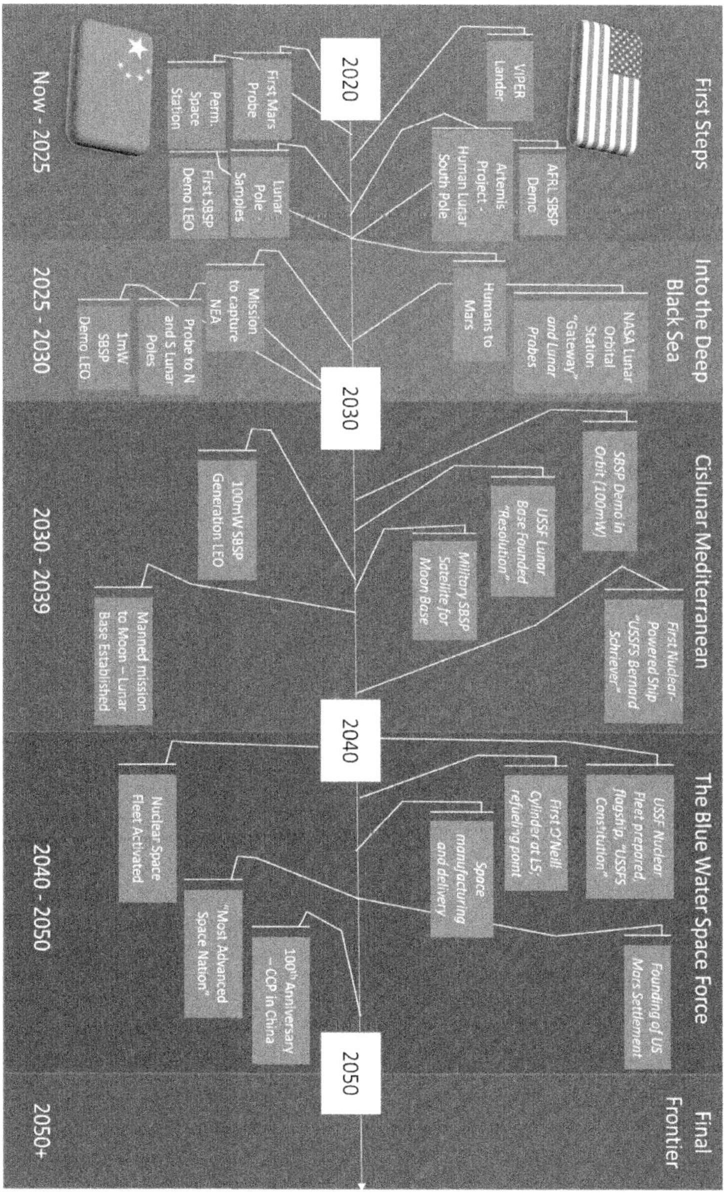

Figure 22 - The New Space Race, Reimagined, The Final Frontier Redux (original by author)

As seen in Figure 22, this second scenario has a distinctly different ending because the strategy is different – the pace of technology development is relatively equal between the two. The US does not get an infusion of technology or some new superweapon; instead, it uses the technology it currently possesses to more strategic ends and out-competes China's attempt to dominate the Silent Sea. The key is moving first to discover strategic locations during *exploration* and then *expanding* to occupy them before hostile nations can do the same. Once complete, *exploit* the resources and allow commerce to do its work. If threatened, *exclude* hostile forces as needed.

The two key events in the timeline are: 1) the founding of the Moonbase and, 2) the nuclear-powered space fleet. If China can achieve these milestones before the US, then the US will almost certainly be in a disadvantageous position. If, on the other hand, the US beats China in both, then the US will likely be able to win the competition and continue to expand beyond Earth's gravity well. Beating China is not the goal, but a benefit, of this strategy. The goal is to *expand* US influence and leadership into space, ensuring future security and prosperity.

It is not technology, individual bravery, or individual industrial creativity that will be decisive in this new space race, but rather a different theory that gives birth to a better strategy. The analysis of the preceding two scenarios shows that the US needs to adopt a more unified and clear strategy – one worthy of the world's superpower – before China can surpass it. More than the future of the world, the future of the Solar System is at stake.

---

[1] Mahan, *The Influence of Sea Power upon History, 1660-1783*, 27.

Chapter 12
Space Development Theory (SDT)

Theory need not be a positive doctrine, a sort of manual for action. It is an analytical investigation leading to a close acquaintance with the subject.

-Carl von Clausewitz

Space Development Theory encompasses the phases of space development, the astronautic/spacepower divide, and a vision for the future blue water spacepower possibilities.

Space Development Theory (SDT), while initially introduced earlier to serve as a framework for the scenarios, is here proposed in its entirety. This theory encompasses all applications of national power in space, posits distinct phases of development, and links it back to the impact that development is likely to have on the nation. Finally, it predicts likely future technologies and developments that could change the nature of national presence in space.

SDT is a framework to understand the purpose of development and systematized it into the phases of space development – the Four E's: *Exploration, Expansion, Exploitation,* and *Exclusion.* As discussed previously, they are generally sequential and progressively build upon each other. The only phase that is likely to occur out of order regularly is E4, *exclusion,* and that is because it is the only phase focused on human competition and the nation's ideational concept of the terrain or area at stake.

Phases of Spacepower – Exploration (E1)

- Entry/Exit Criteria
  - Entry – physical proximity to the location, able to observe, map, and evaluate
  - Actions – identify key locations and resources and phenomena impacting access and exploitation
  - Exit – physical presence and permanent facilities, claims of national interest
- Instruments of National Power:
  - Diplomatic: tasked to gain freedom of action through agreements and restrict competitor freedom of action; benefits alliances through cooperation in the region, convert ideational benefits into restructure of alliances.
  - Information: tasked to project image of international leadership through narrative of vibrant national scientific leadership; benefits from position of international leadership in that area.
  - Military: tasked to assist, or may explore directly; supply, develop, and benefit from the technology established and experience gained.
  - Economic: may serve to stimulate exploration and move to expansion more rapidly if economic benefits of the new region are apparent; evaluates findings and samples for viability of exploitation.
- Motivation for moving to next phase can come from informational, military, or economic vectors and for either positive or negative reasons.

Figure 23 - SDT Phase 1 – Exploration, in detail (original by author)

As seen in Figure 23, *exploration* focuses on gathering information and generally has both scientific and practical applications associated with it. *Exploration* is usually a directed phase, as the nation in question desires to see if there is anything, whether resources or positions, that can be used to its advantage. This phase is the broadest and least-committing. Military forces have historically been used for *exploration,* though not to the same amount recently, because they are domain specialists and are practiced in overcoming resistance in austere environments. For instance, an army is a land domain specialist, whether it is overcoming resistance from a hostile force, or a hostile environment. The same is true of the navy in the sea domain. For instance, the record for the first undersea voyage of the North Pole was by the USS Nautilus, the Navy's first nuclear-powered submarine.[1] Likewise, the most massive naval expedition to the South Pole was the US' Operation Highjump, conducted in 1947.[2] The fledgling US sent a naval expedition into the Atlantic in 1838; after sailing around the tip of South Africa, it continued into the Pacific, where it *explored* Tahiti, Hawaii, and New Zealand.[3] Finally, following hot on the trail of Lewis and Clark, former US Army Major General John C Fremont explored the Rocky Mountains, where he claimed them for the US, and then mapped the Oregon trail, which served as a guide for future immigration.[4]

The military *explored* new regions and mapped them – which is a form of power. As Jason Smith says in his book, *To Master the Boundless Sea,* "[h]istorians of cartography have long understood maps as instruments of power whose boundaries and borders, colors and symbols, and numbers and tables cast a veil of precision and objectivity over the subjective process of constructing meaning, representation, and control." [5] So, while *exploration* seems benign, to fall behind places a nation at a disadvantage in the informational dimension of competition. Possessing the best possible information about an area of a domain, and denying it to the enemy, accrues significant benefit in the next phase, *expansion*.

Phases of Spacepower – Expansion (E2)

- Entry/Exit Criteria
  - Entry – able to sustain presence, either permanent or rotational
  - Actions – establish logistics, observation, and communication outposts, forts, and living areas; either military or economic leading expansion
  - Exit – area in question and logistic lane are secure enough to begin exploitation; net economic benefit to the nation is positive
- Instruments of National Power
  - Diplomatic: tasked to consolidate claims or negotiate non-interference with other powers while attempting to de-legitimize any competing claims; meant to assert ownership of area.
  - Information: tasked to publicly legitimize occupation of the area and emphasize that the occupation presents no threat to the system at large; benefits national power by enabling superior planning in relation to the areas occupied.
  - Military: will likely be required to either occupy or hold the area in opposition to potential rivals, including both force and reconnaissance; benefits from ability to exploit key military terrain in possible conflicts, and gaining experience in the new domain.
  - Economic: commercial interests may take up locations in the new area and then expect military protection; benefits are still small, though niche markets could exist through this expansion.
- Motivation for moving to next phase can come from ideational, commercial, or military vectors, and for either positive or negative reasons.

Figure 24 - SDT Phase 2 – Expansion, in detail (original by author)

As seen in Figure 24, *expansion* is defined as the rooting of the permanent military, commercial, or population entities to establish a permanent presence in an area. *Expansion* employs many tools, but foremost among them is the military tool of national power. Because the new *expansion* will almost certainly be contested if there are either resources or positional advantage to be gained, there is a necessity to defend it. That is what militaries excel at – indeed, projecting national power and defending against hostile projection is the definition of a military's role. Nevertheless, military *expansion* is insufficient, and bases will not grow on their own, as Mahan noted.[6] That being said, the military is critical, as Shufeldt noted, "If the mercantile marine is so essential to the Navy, it is safe to stay the Navy is no less indispensable to commerce. The Navy is, indeed, the pioneer of commerce."[7]

While the military will likely lead, the indispensable growth of commercial and population centers must also be stimulated and encouraged. This growth is, indeed, the rubric used in the USAF's own "Futures of Space 2060" study – evaluating possible scenarios for space development along the axis of population and commerce presence in space.[8] This *expansion* can manifest itself in any number of forms, ranging from trading posts, forts, towns, cities, or other specific entities. Their purpose in the context of SDT remains the same: provide a location for the *expansion* of national power through industry and population growth. Industry and population are significant forces inside states, and a state will only ignore their needs at its peril. This overriding mission underlies the impetus that Mahan describes in his book, where he highlights that a navy that supports commercial shipping will be assured continued funding and life because people and industry require it. The state will respond to the wealthy industry leader's demands. *Expansion* describes what occurred in the lead up to the Mexican-American War, with American settlers moving into Texas, California, and other present-day US States and wresting them away from the Mexican sphere of influence. [9] It would also describe the original British colonies in the Americas, the colonies of the major European powers during the nineteenth century, and the current Chinese plan to create artificial islands to expand their sphere of influence in strategic locations in the SCS.

Phases of Spacepower – Exploitation (E3)

- Entry/Exit Criteria
  - Entry – projected gain is high enough, and risk to commercial interests low enough, for commercial interests to
  - Actions – establish gathering, production and distribution nodes
  - Exit – resource can be exhausted or hostile forces could project sufficient force to cause reevaluation; if promotion, nation considers area key national interest
- Instruments of National Power:
  - Diplomatic: tasked to seek export markets and ensure safety of commerce; benefits distinctly as it can have unique and highly-valuable resources available for leverage.
  - Information: tasked with highlighting benefits of continued national occupation and benefits of trading.
  - Military: tasked with protecting commerce from aggression or predation as well as assisting in construction of infrastructure; benefits from increased funding and a continued clear mission.
  - Economic: economic growth is primary purpose of this phase, must maximize return, construct infrastructure and increase freedom of trade and maneuver; benefits from increased industry, trade, and resource extractions.
- Motivation for moving to next phase can come from ideational or commercial vectors, and generally for positive reasons.

Figure 25 - SDT Phase 3 – Exploitation, in detail (original by author)

Once the *expansion* has taken place, national power grows; the exit criteria is that the new areas begin to generate more wealth than they are taking from the national coffers. In other words, the wealth and population present in the area offset the cost of the military units and infrastructure to ensure continued open lines of communication. The flowing out of national power to secure the area reverses, and now net-benefits start flowing to the nation. That moves *expansion* into *exploitation.*

As seen in Figure 25, *exploitation* primarily benefits a state's economic element of power, bolstering both populations, which aids power through greater wealth generation, and the direct commercial exchange of goods and trade items present, which adds to economic clout. *Exploitation* is defined as a nation's use of a secured area or resource to benefit commercial and population interests to the point that net-benefit accrues to the nation. *Exploitation* is when the people of a state take ownership of *expansion,* and the nation begins to reap the rewards – generally economical, though strategic is possible.

Exploitation can range from the pillaging that the Spanish did in Central and South America to the sustainable colonies that the British set up in America and Canada. It also covers the Straits of Gibraltar and other military outposts that help secure the trade lanes because that is part of securing economic benefit, especially with trade being so crucial to national development.

Phases of Spacepower – Exclusion (E4)

- Entry/Exit Criteria
  - Entry – conditional phase, requires contested control by hostile powers and area viewed ideationally as a member of the national body; may be driven through either political or popular factors
  - Actions – may be offensive or defensive; aims to eliminate hostile occupation or prevent hostile occupation of friendly terrain through denial of occupation, sustainment, or transit; extensive cooperation between military and economic instruments, with military being supported by economic
  - Exit – the contested control is eliminated, either through victory or defeat and focus will shift to a different phase until control is contested again
- Instruments of National Power:
  - Diplomatic: tasked with making hostile action illegitimate and targeting alliances of hostile powers, may seek to gain treaties with friendly or neutral nations.
  - Information: crafting narrative explaining centrality of location to national interest and linking it to national politic.
  - Military: must be able to project sufficient force in the domain to occupy the terrain or be able to deny occupation to hostile powers from nearby domains; primary tool during most exclusion events.
  - Economic: commercial entities are restricted from trade with hostile powers.
- Motivation for moving to another phase is to continue growth of national power; exclusion is by nature a taxing action.

Figure 26 - SDT Phase 4 – Exclusion, in detail (original by author)

As seen in Figure 26, through either political or popular motivation, a nation that feels its control of an area is threatened may embark on *exclusion* to ensure it retains control. This step occurs when a nation has an interest in controlling an area or resource and can mass sufficient force across all elements of national power (diplomatic, information, military, and economic) that it can force non-friendly forces out of the area. *Exclusion* is defined as a nation taking sole ownership of an area or resource and using its massed capability to deter hostile interference to its action. It may be ideational and not purely economic, as a nation may seek to *exclude* a hostile nation from an area with little economic benefit but immense ideational value. This phase is centrally concerned with the opposed nations' relative power and will. Britain's jealous guarding of India would be an example of *exclusion*, as would its continued occupation of "The Rock," Gibraltar. As a specific example in literature, the Russians were threatening India during the 1880-1890s, and Britain played what it termed, "The Great Game" of intrigue to *exclude* Russian designs. The book by Rudyard Kipling, *Kim*, captures this from an adventurous boy's perspective.[10]

    The point of *exclusion* is that it serves the double advantages of securing national sovereignty over the area and can also provide benefits to those viewed favorably by the nation. For example, during war, Britain *excluded* hostile powers attempting to pass through the Straits of

Gibraltar. One specific example occurred during the opening hours of WWI, when Britain captured four German ships attempting to transit.[11]

To provide balance, *exclusion* is not always successful and can be undermined by unanticipated technological capabilities, as the British had significant difficulty stopping German U-Boats from conducting similar transits during the same war. Alternatively, allies like France and America passed through the Straits, a clear example of the diplomatic benefits to being allied with Britain. In this case, in addition to the resource being economic, it became a tool for national strategic benefit.

When a nation's ownership is at stake, it will usually expend significant resources to maintain it. A failed attempt of this was Iraq's invasion of Kuwait in 1990. It occupied Kuwait and immediately began to *exploit* it to rebuild its economy that had been shattered by nearly a decade of war with Iran. It also attempted to *exclude* any other nations from being able to reap the economic benefits of the oil in Kuwait. This *exclusion* was not successful, although the national interest of Iraq was enough to mobilize its military to hold it. It was sufficiently threatening for the US to contest their *exclusion* and mass a coalition to force Iraq from Kuwait. The rest is history, as Iraq's military shattered, and they relinquished Kuwait.

Exploration will evaluate and map a new domain or area. *Expansion* will see some combination of civilian and military elements move in and attempt to permanently occupy critical areas identified during *exploration*. *Exploitation* is when national economies and populations benefit from resources and additional areas for habitation. At any point during the process, with the state's identity entwined with the area or resource, it will attempt to *exclude* hostile actors from interfering in what it considers "its" area. In *excluding,* the military arm will almost always be the primary tool with firm support from the economic. If the hostile power is perceived as having been successfully *excluded*, and no further threats exist, then the nation will likely return to the previously exercised phase of spacepower development to continue to accrue economic benefits.

National instruments of power must be present in a domain to develop it. For the expression of national instruments of power in space, it must be refined into the two mutually supporting categories of spacepower and astronautics to be meaningful. National power in space has traditionally been discussed through the unitary term of *spacepower* or *space power*.[12] As previously mentioned, there is a critical distinction in mature domains between the military and civilian elements that goes unacknowledged in space.

Land power, projected through the army, is as old as humanity itself. Some of the first known recorded histories are of great land battles.[13] Nations on land, delineate between the "army" and "commerce" or "military" and "industry." Without a nation's ability to both generate wealth and field a force to protect it, a nation-state cannot exist.

Seapower likely developed later and was first recognized in the Minoans.[14] Early development came to recognize the mutually supporting but different arenas that came to be known as *seapower* and *maritime power*. *Seapower* is the projected military force capable of action. At the same time, *maritime power* sums all other sea-related capabilities - locations, production, national character, trade and commerce, and overseas basing.[15]

Airpower developed most recently as humans have only been able to fly in heavier-than-air vehicles for about a hundred years. Noted early American airpower apologist, William "Billy" Mitchell, made the delineation between military airpower and what he called "Civil Aviation" in his book.[16] This delineation between *airpower* and *aviation* persists to this day. While Mitchell defined *airpower* as "the ability to do something in the air," it is too broad to be particularly useful as it would encompass both airpower and aviation.[17]

Space needs to recognize the delineation between the military aspects of domain-projection and the non-military aspects. Addressed earlier, *Spacepower* is defined as military force that can exert influence in and from the domain and create effects in other domains for strategic benefit. *Astronautics* is defined as those elements that are primarily commercial and industrial; it includes all aspects that allow for projection into, production, sustainment, training, profit, and expansion in the domain for the purpose of strategic benefit. *Spacepower* is military power in the space domain. *Astronautics* is locational, industrial, and commercial, adding to or enhancing a nation's potential in the domain and provides both motivation for and capability to project military forces both into and from the domain.

This distinction is important because, while the two are mutually supportive, they are not interchangeable. To suggest that warships carry commercial goods as their primary purpose would be just as ridiculous as outfitting commercial vessels with makeshift weaponry to serve as a country's primary navy. Each has distinct, supporting roles to the other. Mahan argued that a navy exists when commercial interests ensure its continued sustainment.[18] If constructed for purely military purposes, "experience [has] showed that his navy was like a growth which having no root soon withers away."[19] As Mahan says that, in general, "[t]he necessity of a navy…springs, therefore, from the existence of a peaceful shipping, and disappears with it."[20]

Merchant ships, meanwhile, sailing far from home do not get far beyond its shores before, "the need is soon felt of points upon which the ships can rely for peaceful trading, for refuge, and supplies." [21] These points are what he calls "colonies," but would be better described as "overseas bases" or "settlements" in modern terms.[22]

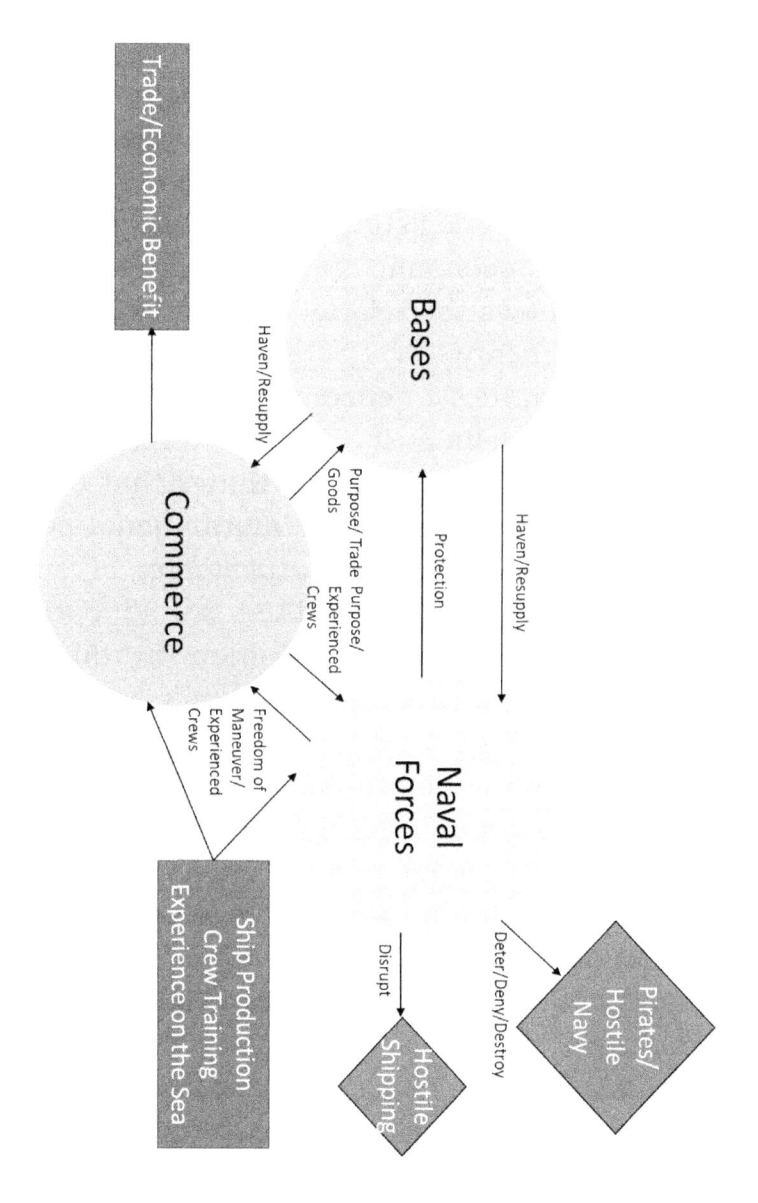

Figure 27 - Basic Diagram of Mahanian Seapower Theory (original by author)

In short, the merchant shipping drives the country's economy and necessitates the other two elements. The navy's need is generated to both protect friendly shipping from hostile or criminal intentions or to project power and impair hostile shipping or naval forces. The bases are needed to provide extended logistics for the maritime forces and amplify economic benefit through trade. Figure 27 shows the relationships between them in Mahan's theory.

Astronautics encompasses Ziarnick's "Grammar" delta. It includes production, shipping, and bases – all the things that bolster the economic, political, and informational power of a nation through the *exploitation* of space resources.[23] SpaceX and NASA would both fit inside the *astronautics* definition in different fashions. SpaceX benefits the nation's production and represents a nascent space shipping interest.[24] NASA adds to the informational power of a nation through continued *exploration* and assists in keeping space production alive. *Astronautics* is the keel of *spacepower*, the foundation upon which *spacepower* is constructed. A poor foundation will lead to a weak structure.

It is possible to generate *spacepower* with a purely military mission, but there are historically only two reasons to do so - either a nation will build a navy to seek national glory or fulfill its aggressive designs.[25] Assuming the nation has no aggressive intentions and is seeking more than recognition through possession of space capabilities, *astronautics* must come first. *Astronautics* must precede *spacepower* and provide a fertile substrate in which spacepower can grow. An example of this being done poorly would be the historical US space mission. Historically, the space race and the Apollo missions were about prestige and, once the US won, there was no impetus to continue.[26] Once again, historically, the *Kriegsmarine* (German Navy) before World War II built some of the most impressive warships in the world. However, they were ultimately bested by the nation that still owned the waves through their superior maritime capability, Great Britain.

Speaking metaphorically, *astronautics* is the tiger and *spacepower* the claws. What then can these claws accomplish if properly sharpened? If seapower helped secure one of the greatest empires the world has ever seen, then *spacepower* can accomplish even more when the 'islands' controlled are planets in the ocean of the Solar System. *Spacepower* is the military capabilities that protect *astronautics, exclude,* or destroy hostile forces and can project effects into other domains – land, sea, air, cyber, and the electromagnetic.

Spacepower comprises all space-based elements of military power, those craft with primarily military utility, or units functioning in a military capacity. Spacepower does not extend to terrestrial-based counter-space capabilities or other non-space-based capabilities. The reason for this is the same reason that surface to air missiles (SAMs) should not considered "airpower." They are another domain's attempt to defend itself from strikes, not a domain capability itself. There is some confusion over this, for instance, India's test of an ASAT and subsequent declaration of itself as a military space power.[27] An ASAT no more qualifies a nation as a spacepower than one possessing SAMs would qualify it as an air power, or one with coastal artillery, a sea power. These are defensive measures meant to deter or threaten inside a limited envelope but do not qualify as domain dominance due to their inflexible nature. For domain dominance, a nation must be able to project sustainable forces into that domain.[28]

    The current small vision of *spacepower* did not have to be. Some early space thinkers in the military looked at developing a Lunar base to serve as a staging ground for future operations.[29] The USAF also considered constructing what would amount to space battleships capable of planetary nuclear bombardment in the event of aggression by the USSR.[30] This dream ended for several reasons, not the least of which was the doctrine of Mutually Assured Destruction (MAD), the policy of both the US and USSR during most of the Cold War. It relied on both nations deterring the other from a first strike through survivable second-strike forces. The Cuban

Missile Crisis, provoked in part by the US nuclear missiles threatening the USSR from Turkey, is an example where both nations attempted to gain advantage and realized that the *status quo* was worth maintaining, given apocalyptic alternatives.[31] In the vision that MAD dictated, space battleships capable of raining down nuclear devastation would certainly count as "destabilizing the *status quo*." The eventual declaration by Ronald Reagan of the Strategic Defense Initiative was a distinct break from previous MAD policy and embraced what was termed "Assured Survival" through the employment of ground-based interceptors and spacepower.[32] This program did not progress much beyond the theoretical stage however, and spacepower emphasis faded. The relegation of spacepower to a purely support domain stunted both development and thought. This marginalization may have continued indefinitely if not for the dramatic explosion of a satellite in orbit on 11 January 2007. On that day, the support domain was a warfighting domain again. It just took a few years for everyone to finally acknowledge it.

On that day, an ASAT missile fired by China struck one of its old satellites, destroying it and spreading a cloud of debris that persists in Earth orbit to this day.[33] While not an act of *spacepower*, it was an explicit threat to deny the domain to the US and removed the feeling of sanctuary that the US enjoyed in space up to that point.[34] Using a WWI analogy, the biplanes flying over the lines on reconnaissance missions were now in danger of being shot down by anti-aircraft guns. Not airpower, but a bold challenge to previously uncontested airpower.

Previous strategies will not work against China – merely attempting to contain them and let them implode from a crippled economy is not viable. China has already surpassed US' purchasing power and, according to projections, will surpass the US in GDP in the next few years. For the first time in living history, the US struggles with an adversary that can match it economically. Also, unlike the USSR, China has been engaged in a far-reaching diplomatic and economic offensive meant to embed them around the world and counter possible isolation strategies. Unless an exceptional threat can be proven, no nation wants to damage its economy by engaging in embargos and other forms of economic punishment against a significant trading partner.

The configuration of US *spacepower* has solidified during decades of organizational reinforcement. Only the combination of the external threat posed by China and the vision of US leaders resulted in the founding of the USSF. Even then, there is a danger of merely replicating the narrow USAF opinions of *spacepower* and missing *astronautics* as a strategic concept entirely. Most importantly, the USSF needs to lead in working to provide a framework for US *astronautics expansion* and *exploitation* of space resources. This growth will enable the US economy to surpass China's again and garner stronger alliances to challenge China's aggression.

There are four possible strategic purposes for *spacepower*. These purposes are consistent with other domains. All of them must: control their domain, strike other domains, conduct reconnaissance of other domains, and move other domain forces through its domain. In this way, *spacepower* has the same missions as air, land, and seapower but executes them differently due to the domain's unique characteristics.[35]

First, and foremost, *spacepower* must always be securing the domain against hostile operations during a conflict – whether war or operations short of war. This grouping covers all awareness, command and control, offensive, and defensive operations inside the domain for deterrence and countering hostile actions. This control encompasses everything from commerce escort and anti-piracy to space military engagements. Securing vital areas in a domain against hostile actions provides freedom of action and maneuver for friendly and allied forces. The British during both World Wars were able to move forces over water with near impunity, even in the teeth of attempted Axis opposition. The US also did so in the Pacific during WWII, moving hundreds of thousands of men thousands of miles in their offensive against Japan.

In short, the primary military purpose in this domain is to secure it, but securing it also benefits the national economic tool, as a safe common is a significant benefit to an open economy like the US. This benefit to the US is why current peer competitors are challenging that openness.[36] Diplomatically, the ability to secure a domain against non-peer threats speaks of national power, and a failure hints at national weakness. One of the first significant campaigns of the US Navy was to crush the Barbary Pirates, which had been plaguing the Europeans for centuries. The US Navy's success helped establish the power of the fledgling nation.[37]

As a subset of *spacepower* securing the domain, it must also provide planetary defense to protect the Earth from the domain, specifically asteroids or meteors. This has been a topic of some discussion in the past and has recently resurfaced.[38] From a humanitarian perspective, there is no USSF mission that can do as much for humanity. Preventing a city-razing or even species-ending asteroid strike on Earth should deservedly garner good will from the other nations protected from the threat. In addition to the benefits to humanity, the organizational benefits of embracing this mission are that it requires space domain awareness and advanced propulsion capabilities that would benefit the USSF in other areas as well.[39]

Second, *spacepower* may eventually have the offensive mission of orbital bombardment, projecting firepower from the space domain into air, sea, or land. This was the intended mission of the Orion project. Whether the method would be a nuclear device, kinetic rods, conventional missiles, lasers, or electronic attack in the EM spectrum, orbital bombardment is the mission of impacting hostile actors in other domains. Ideally, this capability serves to deter threats to national survival due to the threat of destruction held overhead. Unlike seapower, which is physically constrained to the oceans and, therefore, would have difficulty impacting inland areas except through cruise missiles or some similar weapon system, spacepower is literally over everything. There is nowhere it cannot see, and nothing it cannot affect. Unlike airpower, it is persistent through both orbital and deep space systems. This mission set is almost exclusively military in application, though the threat can certainly serve politically and informationally.

This mission, more than any other, raises significant concerns about possible war or even all-out nuclear exchange. The specter of ships armed with weapons capable of reaching the surface in a matter of minutes is undoubtedly jarring to the average person. However, it is unlikely that it would indeed provoke a dramatic reaction if done properly. The USSR and the US both existed for decades under the threat of nuclear bombers, and later both land- and sea-launched nuclear missiles with response times in the minutes, similar to a space-launched weapon, and they did not go to war or cause Armageddon. It is unclear whether the future deployment of space-based ground-attack weapons would follow this same mutually reinforcing deterrence model, but it is possible.

Additionally, even though the OST explicitly forbids weapons of mass destruction in space, the topic is worthy of hypothetical consideration.[40] Treaties are broken, either surreptitiously or explicitly, and the possibility of nuclear-armed spacecraft must be considered when discussing the future due to the dramatic impact they could have. As previously noted, they would not necessarily provoke war. However, any armament in space would need to be handled delicately to avoid projecting an overly hostile intent. Additionally, once national power has projected into space and 3D printing allows the creation of objects without large logistic footprints, confirmation of treaty adherence will be challenging. While nations may wish to close Pandora's box by that point, nuclear or other WMDs would be present and not likely to be discarded willingly. The Pope banned the use of crossbows in Europe but, once a superior weapon emerges, legal restrictions are not effective in removing it.[41] Similarly, in a modern day example, North Korea has refused to give up its nuclear weapons despite stringent embargos.[42] A nation will seek the weaponry that it perceives will assure its survival.

Third, *spacepower* conducts reconnaissance and surveillance missions in support of other domains. This mission is historically one of the primary purposes of spacepower going all the way back to the days of *Corona* in the Cold War.[43] Novel sensors and analysis techniques are continually developed and employed to gather the information needed. Here again, the overhead and persistent nature *spacepower* mean it can establish a nearly constant watch. Reconnaissance serves to bolster all the elements of national power. Knowledge of an enemy's deployments and force locations can assist the military in conducting operations or responding to an emergency. If a surprise attack is feared, confirming hostile deployments can also assist the military in deflecting it. Information and Diplomacy benefit through the gathering and leverage of knowledge gathered through space as well, whether being able to share damaging information about a rival or using that same information to gain diplomatic leverage. Economically, some environmental impacts are visible from space, such as flooding that destroys crops or factories. This knowledge can benefit competitor nations. At a minimum, the economic output of a nation or location could be assessed, or locations of shipments cataloged to allow for counter-economic moves. In all of these, the space reconnaissance mission uses *spacepower* to bolster the other tools of national power.

Fourth, *spacepower* must be able to move forces and resources through the domain, whether that means flying around the Earth or landing on another moon or celestial object entirely. An often-overlooked ability of airpower, bland in comparison to thrilling air-to-air fights and crushing bombing missions, is moving freight and forces around the world as needed. One of the US military's functional commands, TRANSCOM, is explicitly configured to accomplish this task and includes Army, Air Force, and Navy elements.[44]

Logistical readiness has been an effective method of deterrence in the past. For instance, when the USSR attempted to starve out an isolated Berlin from 1948-1949, herculean efforts by the US, British and French aircraft lifted thousands of tons of supplies daily into the besieged city, demonstrating superior will and capability to a stunned USSR.[45] In the future, if a hostile China imposed a sea blockade of Taiwan or some other local nation, spacepower could drop supplies via some delivery method to circumvent their attempted coercion. If a scenario like this occurred, the lift ability of spacepower would need to be significantly more than current capability, through either much larger frames or significantly more of them. While airpower may be able to accomplish a similar feat, if the Chinese emplace lasers, SAMs, or interceptors as part of the blockade, transport aircraft may have difficulty accessing the area.

In the short term, deployment of US forces could occur anywhere in the globe on the order of minutes using space deployment. While not likely able to have enough lift capability to deploy brigades or other large force packages, putting special operations forces from the US into a target location anywhere in the world inside an hour would allow unparalleled flexibility in SOF deployments.[46] Such operations would need to either have local anti-air capabilities suppressed or be absent, as they would likely be unable to defend themselves from modern anti-air missiles as they descended back through the atmosphere.

Looking more long-term, projecting forces from space down through the littoral to the planet would be difficult in the face of dense air defense like what is present in some areas of Earth.[47] However, if the objective is to send forces from space to the Earth, it is not a difficult proposition to simply land around the bubble of SAM protection. This would enable forces to be deployed from space with relative ease almost anywhere on the planet. Most integrated air defenses now are overlapping but primarily planned against air-breathing opponents because that is who they face. Even the best aircraft are unable to break into space and therefore circumvent the air defenses in that way, they must rely on flying low enough to be lost in ground-clutter or stealthy enough to avoid detection. Not so with space forces, which can choose their orbit, descent pattern and other parameters to minimize their vulnerability to air defenses.

  In looking to a future general scenario, blue water spacepower could project capability back to Earth through the littorals. Carrier or battleship-class vessels could begin with standoff from Earth's defenses and engage co-orbital ASATs and other anti-space weaponry emplaced in space if the threat is considered high enough. Once the threat was deemed acceptable, faster cruisers could provide ground bombardment of identified ground ASAT launching areas while landing craft dive through the atmosphere toward their landing zones. Because they are approaching from above, some larger countries like Russia, the US, and China would find it difficult to defend all their land mass through ASAT capability, especially with orbital bombardment causing damage to launchers and radars. While likely not able to provide the sole threat vector unless space industrialization and space personnel was able to project significant forces down into Earth's gravity well, this kind of force presentation would be highly useful for landing special forces at key locations during a conflict.

Significantly easier would be projection through other littorals that are defended much more lightly than Earth. Whether for policing actions, like a "Space Coast Guard," which should be one of the USSF's missions, or deploying military forces through the littoral to another celestial body, the operation would look fairly like the previously described Earth operation. Blue water spacepower would move to the planet and firepower and movement would ensure the landing forces arrive on ground. Less gravity and less defense would likely make this operation significantly less risky than similar ones done against Earth.

Like the discussion on nuclear weapons in space, landing armed forces on another planetary body is explicitly prohibited by the OST.[48] Also like that previous conversation, it is unlikely that the treaty will prevent landing of hostile forces, at least in the long term. Because of the harsh environment of space, much of spacepower will probably use "dual-use" systems to avoid the prohibition against military forces. Dual-use systems can be used for commercial or industrial purposes but also have military applications. Bulldozers and drills, for instance, have applications on Earth to move dirt and put holes in things. However, a bulldozer can also be used for military combat – the Rangers did just that in

Granada.[49] A drill can also be put to more direct applications since it can put holes in anything, including people. The point is that tools usually applied to other tasks can be repurposed for combat with little notice. Therefore, it is not likely that military landings would look like a scene out of some sci-fi movie, with landing craft landing "space marines" in large landing craft like a futuristic Normandy. Instead, it will likely look like large machines capable of industrial labor but also capable of turning their tools from rocks and ice to hostile machines.

    Over time, as the line becomes blurred between industrial and military applications, it is possible that there may be a slow push to militarize the industrial capability of machines in space. This is likely to be covert, at least at first. A historical example of something similar occurred between the World Wars. Under the Treaty of Versailles, Germany was forbidden from producing tanks, so they did anyway as early as 1928, but described them as "light tractors" so that they would not raise suspicion.[50] By the time Germany acknowledged the existence of their tanks and aircraft, both forbidden, it was too late for the Allies to disarm them with anything short of armed conflict. The point is that treaties can be circumvented, and nations will always seek to gain relative advantage. Whether the first military forces on other planets are rapidly repurposed industrial tools or covertly constructed killing machines, humanity will find a way to bring war to the stars.

In summary, Space Development Theory posits that all domain development follows the four phases of *exploration, expansion, exploitation,* and *exclusion,* and that space will be no different. If further argues that space should recognize a similar division between military, *spacepower,* and other power in the domain, *astronautics,* as all other domains do. Finally, there were predictions about the capabilities and missions of future *spacepower.*

For decades, space has been seen through the lens of the Moon landings and science fiction franchises. However, these fail to capture the reality of the tremendous opportunity space presents. For the first time, space *expansion* and *exploitation* will occur during a competition between great powers, and it will almost certainly push both to go faster and further than either would have done independently. In that respect, the competition is perhaps a blessing in disguise, as it provides the impetus for rapid development. Of course, this does not excuse the loss of the conflict. Defeat may still mean dissolution, as the former Soviet Union can attest. It is for this reason that the US and its allies must maintain their position of relative superiority to China, particularly in space. National survival is rooted among the stars.

[1] Danielle DeSimone, "North, South, East, West and Beyond: 5 Military Exploration Expeditions Throughout the Universe", USO, 3 October 2019. (Accessed 10 April 2020: https://www.uso.org/stories/2476-military-exploration-expeditions-through-history)
[2] *Ibid.*
[3] *Ibid.*
[4] *Ibid.*
[5] Jason W. Smith, *To Master the Boundless Sea: The U.S. Navy, the Marine Environment, and the Cartography of Empire.* (United States of America: University of North Carolina Press, 2018), 8.
[6] Mahan, 80.
[7] Robert W. Shufeldt, *The Relation of the Navy to the Commerce of the United States,* Washington, DC: John L. Gink, 1878, 6.
[8] Air Force Space Command, "The Future of Space 2060 and Implications for U.S. Strategy: Report on the Space Futures Workshop." (Washington, DC: US Air Force, 5 September 2019), 6.
[9] Kjetil Ersdal, "Anglo-American colonization in Texas," (Netherlands: University of Groningen). Accessed 10 April 2020, http://www.let.rug.nl/usa/essays/1801-1900/anglo-american-colonization-in-texas/texas-1821-1836.php.
[10] Rudyard Kipling, *Kim.* (London: Wordsworth Editions Limited, 1993).
[11] George Hills, *Rock of Contention: A history of Gibraltar.* (London: Robert Hale & Company, 1974), 398.
[12] Ziarnick, *Developing National Power in Space: A Theoretical Model*, 9.; Edited by Lt Col Kendall K. Brown, *Space Power Integration: Perspectives from Space Weapons Officers* (Montgomery, AL: Air University Press, 2006), 4-10.
[13] Paul K. Davis, *100 Decisive Battles From Ancient Times to the Present* (XX, XX: Oxford University Press, 2001), 1-6.
[14] Mireia Movellan Luis, "Rise and Fall of the Might Minoans" (National Geographic, 2017) Accessed 8 April 2020:

https://www.nationalgeographic.com/history/magazine/2017/09-10/Minoan_Crete/

[15] Mahan, *The Influence of Sea Power upon History, 1660-1783*, 28-89..

[16] William Mitchell and Robert S. Ehlers Jr., *Winged Defense: The Development and Possibilities of Modern Air Power--Economic and Military* (Tuscaloosa, UNITED STATES: University of Alabama Press, 2009), 77–96

[17] Mitchell and Ehlers, *Winged Defense*, xii.

[18] Mahan, *The Influence of Sea Power upon History, 1660-1783*, 48.

[19] *Ibid*, 88.

[20] *Ibid*, 26.

[21] *Ibid*, 27.

[22] *Ibid*, 49.

[23] Ziarnick, *Developing National Power in Space: A Theoretical Model*, 16.

[24] Mike Wall, "SpaceX's Starship will soon be made of different stuff," (Space.com, 12 March 2020.) Accessed 15 April 2020: https://www.space.com/spacex-starship-new-stainless-steel-alloy.html.

[25] Mahan, *The Influence of Sea Power upon History, 1660-1783*, 26; Andrew Lambert, *Seapower States: Maritime Culture, Continental Empires and the Conflict That Made the Modern World* (New Haven, CT: YUP New Haven and London, 2018), 264–65.

[26] Dennis Wingo, "The Early Space Age, The Path Not Taken Then, But Now? (Part II)" Wordpress.com, 17 February 2015. Accessed 15 April 2020: https://denniswingo.wordpress.com/2015/02/17/the-early-space-age-the-path-not-taken-then-but-now-part-ii/.

[27] Sanjeev Miglani and Krishna N. Das, "Modi hails India as military space power after anti-satellite missile test" (Reuters, 27 March 2019) Accessed 15 April 2020: https://www.reuters.com/article/us-india-satellite/modi-hails-india-as-military-space-power-after-anti-satellite-missile-test-idUSKCN1R80IA

[28] Peter Garretson, "USAF Strategic Development of a Domain," Over the Horizon Journal, 10 July 2017.

https://othjournal.com/2017/07/10/strategic-domain-development/

[29] United States Air Force. Project Horizon Volume I: Summary and Supporting Considerations. Washington, DC: United States Air Force, 1959. Document is now declassified.

[30] George Dyson, *Project Orion: The True Story of the Atomic Spaceship* (New York, NY: Henry Holt and Company, 2002.) 214.

[31] Graham Allison and Philip Zelikow, *Essence of Decision: Explaining the Cuban Missile Crisis*, Second Edition (New York: Longman, 1999), 77–99.

[32] Ben Bova, *Star Peace: Assured Survival*. Tor Book; New York, New York. Reprint August 1986.

[33] Brian Weeden, "2007 Chinese Anti-Satellite Test Fact Sheet," Secure World Foundation, 23 November 2010. Accessed 29 May 2020: https://swfound.org/media/9550/chinese_asat_fact_sheet_updated_2012.pdf.

[34] Brian Weeden, "The End of Sanctuary in Space: Why America is considering getting more aggressive in orbit," Medium.com, 7 January 2015. Accessed 11 June 2020: https://medium.com/war-is-boring/the-end-of-sanctuary-in-space-2d58fba741a

[35] For a further discussion of the Space Force Missions, see Dustin Grant and Matthew Neil, *The Case for Space: A Legislative Framework for an Independent United States Space Force*, Air University Press, Maxwell AFB, February 2020.

[36] Joint Chiefs of Staff, *Joint Operating Environment 2035 (JOE 2035): The Joint Force in a Contested and Disordered World*, 2016, 30–32.

[37] The Editors of the Encyclopedia Britannica, "First Barbary War." (Encyclopedia Britannica, Inc., 7 May 2019) Accessed 15 April 2020: https://www.britannica.com/event/First-Barbary-War.

[38] Peter Garretson, "Make planetary defense a Space Force mission," The Hill, 1 June 2020. Accessed 10 June 2020: https://thehill.com/opinion/national-security/500457-make-planetary-defense-a-space-force-mission.

[39] Garretson, "Planetary Defense"
[40] United Nations, "Treaty on Principles Governing the Activities of States in the Exploration and Use of Outer Space, including the Moon and Other Celestial Bodies," UN Resolution 2222, 19 December 1966, Article IV.
[41] Council Fathers, Second Lateran Council, 1139 CE, Pronouncement 29. Accessed 2 June 2020: https://www.papalencyclicals.net/Councils/ecum10.htm.
[42] Kim Tong-Hyung, "North Korea's Kim touts strategic weapon amid stall in talks," AP, 1 January 2020. Accessed 12 June 2020: https://apnews.com/3262c51730fe25feb2a81f036971a56b.
[43] CIA, "CORONA: Declassified" CIA New & Information, 15 February 2015. Accessed 15 April 2020: https://www.cia.gov/news-information/featured-story-archive/2015-featured-story-archive/corona-declassified.html
[44] United States Air Force, "About USTRANSCOM" Accessed 22 April 2020: https://www.ustranscom.mil/cmd/aboutustc.cfm;
[45] History.com Editors, "Berlin Airlift" (History.com, 9 March 2011). Accessed 22 April 2020: https://www.history.com/topics/cold-war/berlin-airlift
[46] Mike Brown, "SpaceX Starthip: Those 30-Minute, Cross-Planet Flights Will Be Punishing," Inverse.com, 27 June 2019. Accessed 2 June 2020: https://www.inverse.com/article/57135-spacex-starship-rockets-will-get-you-from-paris-to-nyc-in-30-minutes.
[47] John Pike, "KPAF – Air Defense," GlobalSecurity.org, 26 December 2016. Accessed 12 June 2020: https://www.globalsecurity.org/military/world/dprk/air-force-air-defense.htm.
[48] United Nations, *Treaty on Principles Governing the Activities of States in the Exploration and Use of Outer Space, including the Moon and Other Celestial Bodies,* adopted 19 December 1966, Article IV.
[49] Logan Nye, "That time Rangers stole a bulldozer for an assault vehicle," We Are The Mighty.com, 5 April, 2019. Accessed 12 June 2020:

https://www.wearethemighty.com/history/point-salines-bulldozer-stolen-rangers?rebelltitem=1#rebelltitem1.

[50] John Pike, "Leichttraktor (VK-31)," GlobalSecurity.org, 24 November 2018. Accessed 12 June 2020: https://www.globalsecurity.org/military/world/europe/history/de-leihttraktor.htm

Chapter 13
Meeting the Future

Never let the future disturb you. You will meet it, if you have to, with the same weapons of reason which today arm you against the present.

-Marcus Aurelius

After astronautics and spacepower have *expanded* into cislunar space, what next? This chapter grapples with likely future tech, missions, and key resources to be found in space. It is meant to be an exploration of the possible, not necessarily a prediction of what will happen.

It is always tricky to make predictions, especially about the future.[1] An unforeseen disruptive technology or event may completely overturn the most carefully crafted prognostications. However, it seems possible to make a few estimations based on current resources, trajectories, and human nature. Assuming a continual *expansion* of human presence in space for gathering and processing resources, like previous national competitions that have occurred, then it is possible to make a few predictions with some level of confidence.

First, the *expansion* will be to capture critical resources, such as minerals on the Moon or Mars, asteroids, or He^3 from Uranus.[2] Working with the assumption of larger ships and more efficient in-space production, there is little reason to doubt that AI-enabled harvesting of resources in the Solar System will be both possible and profitable. This *exploitation* would fuel Earth's economy with minerals and other raw materials to a level unparalleled in its history. According to some estimates, if all He³ were extracted from Uranus, it would provide sufficient energy to the Earth for four billion years.[3] These in-space resources will likely serve as the cornerstone of a space manufacturing industry.

Second, the *expansion* will encourage human habitation in space. Whether that is Lunar or Martian settlement, or would take the form of O'Neill cylinders configured to the desired gravity and climate, this habitation will occur similar to the other significant *expansions* in history.[4] Estimates are that, using the resources from asteroids, there is sufficient living area for trillions of people to live in space.[5] The outcasts, the desperate, and the adventurers will leave the fringes of society and try their luck among the stars. Militaries will explore and hold terrain, while commercial interests will generate money from it. The test of national power will come when these settlers encounter violence, either from pirates, thieves, or hostile settlers. If national power cannot project to protect these new areas, another nation, that can do so, will add the unsupported settlements to its domain.

The third point is that war will occur in space. The current mantra is that space is a warfighting domain. However, the future space domain will be more than satellite orbits and concerns about debris fields. With nuclear-power fleets likely by 2050, space warfare will see true maneuver, firepower, and communication. [6] Current space doctrine attempts to avoid debris because orbits can be lost or severely degraded from waste orbiting a planet. In contrast, in deep space, the amount of distance between objects is, literally, astronomical. Therefore, debris is significantly less of a concern as even large clouds of it would be lost in the pure scope under discussion.

On a side note, even if combat occurred in a planetary littoral region, there are possible mitigations. There are still feasible technical methods to remove debris from orbit, and some recent articles recommend this mission be given to the USSF.[7] Methods include deorbiting smaller pieces with particle beams or larger ones with electromagnetic tethers or robotic servicers. There is opportunity for building synergistic international cooperation as the current legal framework does not incentivize space salvage or debris clearing. [8] Renegotiating these to encourage space salvage would be a valuable step forward for both astronautics and international cooperation.

In deep space, however, combat will be far different from anything in orbit – as the distances will be significant and debris no concern. In deep space, it is unclear what precise weapons such ships will have, but they will probably fall into similar categories to the current naval weaponry categories of guided projectiles, direct-fire weapons, and defensive weaponry. Also, likely to be present is some adaptation of current naval aviation.

Guided projectiles are currently divided into sub-surface torpedoes and air-breathing missiles. Missiles are subdivided into whether they attack naval, air, or land targets. In future space warfare, there will likely be some delineation between atmospheric and void weaponry. Weapons flying through the atmosphere will require heat shielding to avoid incineration, while weapons hurling through the void will need radiation shielding. Additionally, the intended target for these weapons would also cause a branch inside this category. Speed and maneuverability will undoubtedly be needed for counter-space weapons to ensure they can strike rapidly maneuvering spaceships.

In contrast, speed is the only likely requirement for orbital-bombardment focused weaponry. Weapons capable of evasive maneuvers to avoid anti-missile weapons are likely, but maneuverability requirements will probably not be on the same order of magnitude as anti-ship weaponry because of the spaces involved. Warheads for deep-space applications will probably be either nuclear, neutron, or some other kind of high-yield explosives as the ability to inflict damage will be at a premium. Nuclear has the advantage of being the highest-yield available to most modern warheads. Even near-hits with either a nuclear or neutron weapon would likely cause such suffering to the crew that the fight may conclude after one exchange. This selection of weaponry would also be successful at targeting the infrastructure of ground-based targets.

Direct-fire weaponry would be some form of unguided projectiles meant to damage or destroy other objects. Naval guns, whether the automatic weapons meant to dissuade pirates or the awesome main cannons on a battleship, are contained in the naval equivalent to this category. The space permutation of this could take the form of railgun projectiles, lasers, or some other futuristic system. In space, lasers and other directed-energy weapons do not degrade as they do in the atmosphere, though they still suffer R^2 loss. Therefore, they have significantly less diffraction and would likely have far greater range and capability in space. For anti-ship or orbital bombardment missions, tungsten rods fired from railguns or some other cannon could have devastating effects.

Defensive weaponry covers the full spectrum of electromagnetic jammers, dazzlers, point defense weapons, and emergency thrusters. Sensors in space will need to gather vast amounts of information to sort out hostile ships against a complex background, likely by sorting through ship-generated emissions in the EM spectrum. By attempting to flood the EM spectrum with energy, it is possible to jam enemy sensors and incoming weapons by overwhelming them. Likewise, electro-optical or heat-seeking warheads could be targeted by lasers to "dazzle" them and possibly destroy them. Point defense, like the modern "Phalanx" weapon system found on many US Navy ships, is designed to meet closing missiles with a storm of steel. While not a guarantee of avoiding missile damage, it can significantly reduce the number of hits sustained. Finally, unlike naval warfare, evasive maneuvers are a real element of space warfare. Therefore, emergency thrusters would be considered defensive weapons as they could push a ship out of the way of incoming direct-fire weapons or help buy time for point-defense to engage a few more incoming missiles.

The combination of direct-fire and guided weapons are what makes Earth-based defensive strategies relatively worthless against a space-based threat. Earth is a moving target, but a predictable moving target. A spaceship is not. Assuming a spaceship started at Mars and began an attack run on Earth, or any nation on Earth, almost every advantage accrues to that spaceship. It can choose its course, speed, and maneuver at leisure. If it were targeting a specific nation, it could time the weapon arrival to the Earth's rotation, so the right nation is on the approached side of the globe. Firing its missiles, the ship can begin evasive maneuvers – any counter-shot by the Earth-based defender may find itself thousands of kilometers away from its maneuvering target.

In contrast, the Earth will continue to move and rotate predictably. The nation must rely on interceptors to attempt to stop the incoming missiles. If the attacking ship uses nuclear warheads, they must intercept even farther out to avoid the electromagnetic pulse (EMP). The ship's missiles can maneuver, making this an exceedingly tricky shot at best as the interceptors struggle to climb out of Earth's gravity well. If even one nuclear warhead reaches the right orbital height over the desired nation and detonates, the nation goes dark. Power, computers, vehicles, medical facilities, if unhardened, all gone in a blink.

What could be termed 'space strategic attack' is many times more effective than attempting to launch many missiles from Earth-based facilities. These are known and watched by the growing space intelligence constellations of Earth nations today. A ground launch of rockets carrying nuclear payloads would provoke a response within minutes, if not seconds. Space is vast, and a space-based attack run would be nearly impossible to anticipate. NASA is still occasionally surprised by asteroids passing within 73,000 miles (1/3 the distance from Earth to the Moon), and these are objects with non-evasive movement.[9] Space awareness sensors would need significant upgrades to have any hope of anticipating space-based strategic attacks. Such an attack would be a first-strike capability second to none.

For a more tactical application, such as precise destruction of facilities or forces on the ground without shattering a nation, the future ships could use their direct-fire weaponry. Once again, the Earth's predictable movement and rotation mean that shots from space are relatively accurate and precise. A ship maneuvering at low Earth orbit (LEO) would be somewhere between 99-1200 miles above the surface of the Earth.[10] Some of the best railguns under testing by the US Navy can launch a projectile at Mach 7 (5,370 mph).[11] If the ship providing fire were sitting at orbit at 120 miles above the surface, its railguns would reach the surface in just one minute and thirty-four seconds. This calculation assumes constant projectile speed between the competing forces of gravity and friction.[12] By comparison, traditional close air support may require repositioning of aircraft, possibly in contested airspace, to achieve effects on the target. This vectoring could take minutes to hours, depending on the distance. A space cruiser can see and engage targets in different countries instantly because of its higher altitude.

Finally, there appears to be a role for small, AI-driven craft that would function like fighters and bombers in a modern naval warfare context – in both space and space-littoral warfare, possibly even expanding into the atmospheric. Assuming manufacture capabilities in the "carrier" ships that would allow both the refining and use of resources from space, and some specialized ship to assist in gathering the raw resources – a carrier-class vessel would be able to operate with near-impunity from the blue water of space while controlling deep space, the littorals, and even some of the air domain.

The basic configuration of the fighters and bombers in question assumes it is AI-driven – autonomous, semi-autonomous or swarm in construct. It must be AI-driven because the fighters will be maneuvering at g-forces that would turn humans into jelly, and therefore human occupation would prove to be a detriment. If the spacecraft are autonomously configured, they would be given a mission like a human pilot and then attempt to carry it out as best as possible with no further interaction aside from possible re-tasking. This would require a high level of AI. Semi-autonomous would allow for some human piloting, like many drones now used in air warfare, but would automate some systems or allow for some basic non-human control to allow for lag. If the control lapsed, they could be set to return to the ship on a straight course. In a swarm construct, the massed AI would allow many smaller drones to operate as a greater whole, massing weapons and processing power. Dispersed in this fashion, they could survive hostile fire and, through the massed processing power, operate as either autonomous or semi-autonomous drones.

Engines and propellant on these fighters would be at a premium, so either they will have very efficient drives capable of long burns for speed and fast burns for evasive maneuvers, or they will require some form of solar or beamed power from the carrier. If offensive, the fighters would likely use either solar or internal propellant to allow for longer ranges, moving sufficiently far away from their carrier to engage and destroy the hostile ship while the carrier stays out of weapon range. If defensive, possibly operating as a missile shield of mini-phalanx craft, then beamed power is definitely viable as the craft in question would not need to stray far from the carrier they are protecting.

Weaponry on these fighters could range from miniature lasers to some form of ballistic projectile. Speed in space warfare is significantly higher than any other form of warfare, with velocities measured in kilometers per second – and the speed of craft adding to the speed of any projectile launched. For instance, current estimates for minimal speed for a satellite to stay in low earth orbit is 7.79 kilometers per second, or approximately 25,500 feet per second.[13] For comparison, the muzzle velocity of a 5.56mm NATO round fired from an M4, used by the US military, is 3,250 feet per second, one-seventh the speed.[14] Additionally, the speeds are additive – the resulting speed of a 5.56mm fired from a satellite in LEO would be approximately 28,750 feet per second.

Because mass and velocity are multiplicative (momentum = mass x velocity), the mass of any projectile fired at these speeds could be small, and no amount of armor would protect against it. The kinetic penetrator that the M1A1 uses, for instance, weighs about nine pounds and has a muzzle velocity of approximately 5,700 feet per second. Once again, the satellite itself is already moving five times faster than the weapon that has proved able to destroy some of the most heavily protected systems in the world, tanks. All of this is to say that space warfare could definitely see masses of relatively cheap, agile, and lethal fighters using either laser or kinetic weaponry to destroy hostile fighters or even larger ships due to the difficulty in protecting against that weaponry moving at such high speeds.

Fighter-carried missiles would also be possible, especially if they housed flechette warheads to pepper enemy ships with hundreds of small projectiles. These could cause depressurization in the hostile ship as hundreds of holes are punched in the skin of the vessel. Against hostile fighters, the darts would likely destroy it completely due to the combined speed of the fighter and missile closing, assuming a head-on detonation.

There are applications in littoral warfare as well, and here the carriers' fighter compliment could prove to be highly dangerous to most satellite systems. Assuming the capabilities already described for deep space warfare, fighters with these capabilities would be able to maneuver toward a planet with known satellite trajectories and engage specific objects as desired. Because the space fighters would be moving so quickly, and able to maneuver, ground launched ASATs would be minimally effective while co-orbital ASATs would likely be ineffective. Since the fighters are not co-orbital, but rather extra-orbital, it is unclear whether they could even be targeted.

Of course, deterrence and coercion require that force be threatened but not used. Simply positioning one of these carriers in cislunar space would undoubtedly send a message no less clear than the modern US carrier group operating in potentially hostile water. All satellites in orbit would be simultaneously threatened with destruction in one form or another, and any space-going commerce would be similarly held at risk. The carrier has been the naval power-projection platform *par excellence* since World War II, and the space carrier will likely continue to prove to be so for the foreseeable future.

Depending on the design, these fighters could also operate in the atmosphere – though here they would likely need to possess some "look down, shoot down" capability to engage their air-breathing counterparts below them. Because they are designed primarily for space warfare and not aerial, it would be silly to propose that they could conduct dog fighting with their aerial brethren. Firing missiles from extreme high altitude, likely with autonomous warheads to seek the movement or thermal plume of aerial maneuver, they could assist terrestrial allies by removing much of the air cover over an area of operations while not placing themselves at risk to anything but the most agile ASATs.

Because the weaponry of space does not allow, at least with current and likely near-future technology, an adequate defense against the weaponry likely to be encountered, space fighters will likely have a role in space combat. From an economic perspective, as well as a human, losing twenty relatively cheap fighters for one 'cruiser' or 'battleship' is a distinctly good trade. Losing zero humans, due to AI-flown fighters, compared to whatever was required to crew the bigger ship is, also, a particularly good trade in terms of human life. Additionally, a mass of fighters can each suffer damage individually and they suffer only subtractive damage – that is, each fighter is destroyed and removed separately. Larger ships, while they have redundant systems, suffer divisional damage – that is, as the ship takes damage, it may lose whole systems or capabilities due to its integrated nature. A lucky hit to a reactor, for instance, could provoke a ship-ending explosion; no such strike is capable on a mass of fighters. Sufficiently spread out and hardened, even a nuclear explosion may not be enough to destroy them in one stroke.

These are predictions for future spaceship weaponry, based on current trends and known future technology under development. While Earth-based defensive measures will almost certainly improve as space warfare develops, the new capability presented by this domain will likely change the traditional military calculus. These ships, whether crewed or automated, will serve as the power projection platforms of the twenty-second century.

Fourth, this *expansion* will be aided or hindered by the possession of chokepoints such as Lunar and Martian bases, Lagrange Points, asteroid tracks, and by creating artificial islands in space to serve as servicing locations. This is in keeping with Mahan's assertions about the importance of bases in developing and expanding seapower more than 130 years ago. [15] Many powers have used small bases on a global scale to enhance or enable military force projection, and space will be no different.

Fifth, and finally, there will come a day when humanity will escape the gravity of our sun, Sol, completely, and venture beyond this Solar System. By that point, likely centuries in the future, humanity will have spread throughout the various planets described and possibly several moons or other bodies. Perhaps it will become a utopia like some optimistically present. There may be unity among humanity as conflict ceases, and they put all their will toward *exploration*. That may someday occur, though the chance currently seems incredibly remote. More likely than not, humanity will continue in its struggle between groups that vie for power and unite around common cultural themes such as beliefs, appearance, or language. Since this second situation is much more likely and history warns of the dangers of becoming a second-rate power in the continual struggle, it behooves the US to act now to ensure continued success in the future.

While that struggle continues, the accumulation and use of both economic and military power will prove crucial. This necessity has always been the case - from the team warfare tactics employed by the ancient Assyrian military to the applied science of English longbows that allowed the British to dominate the European battlefield for two centuries.[16] The ironclad, the steamship, artillery, aircraft, satellites. The world is rife with conflict, and humanity is, and always will be, the fiercest competitors. As humanity prepares to step beyond the shores of our island, it will bring this competition with it. Wars will mar the surface of planets not yet visited. Because this is the case, it is crucial to be prepared for this and amass both applied science and the mental map to use those tools in the most efficient way possible, which is tactics.

No society in the history of the world has ever endured. Every empire has eventually collapsed under its weight, and America will be no different. Nevertheless, that day is not here yet. China is a threat, but not every rising power ascends. Instead, as we look to the future and the stars, let us echo one of history's most pugnacious competitors, Winston Churchill, in saying, "Let us therefore brace ourselves to our duties, and so bear ourselves, that if [America]…last for a thousand years, men will still say, 'This was their finest hour.'"[17]

---

[1] Quote Investigator, "It's Difficult to Make Predictions, Especially About the Future." Accessed 17 June 2020: https://quoteinvestigator.com/2013/10/20/no-predict/.
[2] John S. Lewis, *Mining the Sky: Unfold Riches of Asteroids, Comets, and Planets.* (Helix Books, New York, New York, 1997), 204-213.

[3] Lewis, *Mining the Sky*, 211.
[4] Gerard K. O'Neill, *The High Frontier: Human Colonies in Space* (Canada: Apogee Books, 2000), 37-44.
[5] Robert Walker, *Asteroid Resources Could Create Space Habs For Trillions; Land Area of a Thousand Earths*, Kindle Publishing, 8 September 2015.
https://www.amazon.com/Asteroid-Resources-Could-Create-Trillions-ebook/dp/B0156KQ3VS.
[6] Steven Kwast, "The Urgent Need for a U.S. Space Force" (lecture, Hillsdale College, Hillsdale, MI, 16 December 2019).
[7] Roger X. Leonard, "A treatise on the formation of a US space force" The Space Review, 22 January 2018. Accessed 8 June 2020:
https://www.thespacereview.com/article/3411/1.
[8] Peter Garretson, Alfred B. Anzaldua, and Hoyd Davidson, "Catalyzing space debris removal, salvage, and use," The Space Review, 9 December 2019. Accessed 8 June 2020:
https://www.thespacereview.com/article/3847/1.
[9] Allyson Chiu, "'It snuck up on us': Scientists stunned by 'city-killer' asteroid that just missed Earth" (The Washington Post, 26 July 2019.) Accessed 22 April 2020:
https://www.washingtonpost.com/nation/2019/07/26/it-snuck-up-us-city-killer-asteroid-just-missed-earth-scientists-almost-didnt-detect-it-time/
[10] Matt Williams, "What is Low Earth Orbit?" Universe Today: 6 January 2017. Accessed 22 April 2020:
https://www.universetoday.com/85322/what-is-low-earth-orbit/
[11] Allen McDouffee, "Navy's New Railgun Can Hurl a Shell Over 5,000 MPH." (Wired.com, 9 April 2014). Accessed 22 April 2020:
https://www.wired.com/2014/04/electromagnetic-railgun-launcher/
[12] (120/5370) x 60 = 1.34 minutes
[13] Australian Space Academy, "Orbital Parameters: Low Earth Circular Orbits," Australian Space Academy. Accessed 10 June 2020:
http://www.spaceacademy.net.au/watch/track/leopars.htm.

[14] Bill Marr, "223 Remington/5.56mm NATO barrel length and velocity: 26 inches to 6 inches," Rifleshooter.com, 7 December 2015. Accessed 10 June 2020:
https://rifleshooter.com/2015/12/223-remington-5-56mm-nato-barrel-length-and-velocity-26-inches-to-6-inches/.
[15] Mahan, *The Influence of Sea Power upon History, 1660-1783*, 49.
[16] Paul Kriwaczek, *Babylon: Mesopotamia and the Birth of Civilization,* Thomas Dunne Books, 2010, 236;
Christopher Rothero, *The Scottish and Welsh wars, 1250-1400.* Men at Arms, London: Osprey, 1984, 4.
[17] "If the Empire lasts for a thousand years men will say, this was their finest hour." *The War Illustrated*, 28 June 1940. Accessed 29 April 2020:
http://www.ibiblio.org/pha/TWI/TWI-40-06-28.pdf.

Chapter 14
Conclusion

A good Navy is not a provocation to war. It is the surest guaranty of peace.

-President Theodore Roosevelt

True to the chapter title, this is the conclusion where we...conclude.

This book touched on the current US and Chinese space theories and strategies before comparing both through two scenarios extrapolating each over the next thirty years. China's current theory and strategy were shown to be superior to those of the US – which means the US needs to reevaluate how it thinks about space. Because the current strategy will not work, a new recommended space strategy was examined in some detail and assessed to bear better results than the current US viewpoint. This new strategy was also put through the same simulation with the differences highlighted. Based on the examination, the latter strategy was successful. SDT is the summary of the space development framework and incorporates lessons learned from the two scenarios to assist future theorists to understand space.

The underlying prescription is to grapple with reality and seek the most effective ways to live in concert with it. Theory attempts to describe reality and humanity's relation to it in terms that are both helpful and truthful. While theories may, and often are, different, the reality remains consistent – therefore, there is only one actual reality, and humanity attempts to grapple with it as best it can. All theories and models are wrong, but some are useful. Said another way, humanity has opinions about the reality and attempts to correspond but will always only partially grasp it. Humility in seeking is, therefore, crucial in the quest, because any theory, however well-constructed, will fall short of fully capturing the complexity of reality.

While it is prevalent to speak of 'your truth' and 'my truth' this is not a helpful mindset at all. Opinions and theories will always lack perfection because fallible beings make them, but that does not mean that some are not more accurate than others. The more theory or opinion corresponds with reality, the more beneficial that belief is to the person, or civilization, in question. That is why the search for truth and reality must be ceaseless and exacting. Not everything about a differing opinion is right, but something is worth the other person believing it. Refusing interaction makes our minds smaller as humans.

This mindset gave birth to Space Development Theory. While not perfect, it is an attempt to grapple with the reality of space, the opportunities it presents, and how to go about realizing them. Hopefully, this will spur a dialogue about the future. Changes or updates that make a theory correspond more with reality should always be welcomed.

In closing, the focus on the future of space should not be between what technologies to pursue, but rather which strategy to choose. The US must think beyond satellites and footprints on the Moon or Mars. The development process cannot stop at *exploration*; it must be followed by *expansion* and *exploitation*. To stop at *exploration* is to stall development entirely.

A nation can only ascend as far as its vision will take it. Once the people believe that there is nothing further, then the country will fall into decadence and slouch into destruction. Where there is no vision, there is no future. The USSF, more than any other military service, is the guardian of this vision because space has always been the final dream of humanity. The primary threat to this dream comes from the US' primary competitor, China, which has declared its intention to dominate space. Two major scenarios for the future exist, one with a victorious China and one where the US retains its dominant position in the world. Now is the time to solidify the USSF vision and unite the nation behind it. Like the great nations of old, the US must expand beyond its shores for both economic and military power. However, unlike earlier times, it is not the ocean beckoning. The Space Force, and eventually other friendly space forces, are the guardians of our political future. They will be the Sentinels of the Silent Sea.

Index

3D printers, 76, 78
3D printing, 74, 76, 105, 149, 169, 200
Africa, 23, 49, 129, 177
aircraft, 2, 5, 32, 42, 34, 90, 111, 196, 202, 204, 206, 220, 228
Alfred Thayer Mahan, 3, 5, 3, 20, 20, 54, 159, 173, 180, 181, 189, 190, 192, 208, 209, 226, 230
artificial intelligence, 34, 76, 77, 78, 84, 105, 106, 214, 221, 222, 225
ASAT, 2, 42, 44, 110, 111, 136, 194, 195, 204, 224, 225
astronautic, 94, 147, 150, 151, 166, 174
astronautics, 10, 11, 27, 28, 32, 34, 36, 39, 44, 49, 58, 59, 70, 105, 109, 123, 134, 140, 151, 152, 165, 187, 192, 193, 196, 207, 213, 215
Astronautics, 8, 27, 28, 30, 38, 134, 152, 189, 192, 193
Astropolitik, *37*, 56, 57, 131
basing, 21, 30, 36, 143, 146, 147, 151, 168, 188
Blue Origin, 34

blue water, 43
Bradley Townsend, 6, 56, 59
Brent Ziarnick, 6, 27, 27, 55, 56, 63, 131, 155
Britain, 20, 23, 185, 186, 193
British, 19, 20, 21, 111, 135, 181, 183, 186, 197, 202, 228
Carl von Clausewitz, 58, 63, 94, 129, 174
Chang'e 6, 69, 98
Chang'e-4, 2
Chess, 4, 46, 48
China, 1, 2, 3, 9, 10, 1, 2, 3, 5, 6, 8, 2, 3, 4, 5, 6, 8, 8, 13, *14*, 14, 15, 46, 48, 49, 50, 51, 52, 53, 54, 56, 60, 64, 68, 69, 70, 71, 72, 73, 74, 75, 76, 77, 78, 79, 80, 81, 82, 83, 84, 85, 86, 87, 90, 93, 96, 97, 98, 99, 101, 105, 106, 108, 109, 111, 112, 113, 114, 116, 128, 129, 130, 133, 139, 140, 141, 143, 144, 146, 147, 149, 151, 152, 153, 154, 156, 159, 162, 163, 164, 166, 168, 169, 172, 173, 195, 196, 202, 204, 207, 228, 231, 233

Chinese, 2, *3*, 4, 10, *4*,
    5, *3*, 46, 47, 48, 49, 50,
    51, 52, 55, 56, 68, 69, 72,
    74, 75, 79, 81, 82, 84, 88,
    90, 95, 98, 100, 103, 104,
    107, 109, 111, 112, 113,
    114, 134, 139, 140, 141,
    181, 203, 210, 231
Chinese space theory,
    47, 51, 52, 55
cislunar, 22, 60, 98, 101,
    104, 105, 114, 118, 135,
    147, 153, 159, 162, 164,
    165, 166, 213, 224
Cold War, 1, 3, 5, 11, *1*,
    43, 56, 63, 73, 88, 152,
    194, 201
commerce, 4, 30, 31,
    35, 39, 137, 152, 166,
    172, 180, 188, 197, 225
commercial, 4, 3, 6, 28,
    32, *33*, 40, 57, 59, 69,
    109, 114, 137, 139, 144,
    165, 179, 180, 183, 189,
    205, 214
Confucius Institute, 79
Cuban Missile Crisis,
    195, 210
diplomatic, 2, 17, 18,
    21, 71, 128, 185, 186,
    196, 201
Dr. Robert Goddard,
    30
Dr. Wernher von
    Braun, 31

Earth, 2, *3*, 5, 9, 10, 6,
    7, *2*, *15*, 22, 25, 31,
    33, 34, 36, 37, 38,
    42, 44, 50, 54, 55, 57,
    60, 69, 76, 77, 82, 87, 91,
    96, 98, 101, 102, 103,
    105, 110, 111, 113, 116,
    117, 119, 120, 121, 122,
    123, 124, 125, 126, 127,
    128, 129, 130, 134, 136,
    139, 149, 153, 154, 159,
    162, 163, 165, 166, 169,
    172, 195, 198, 202, 203,
    204, 205, 214, 218, 219,
    220, 226, 229
economic, 2, 5, *2*, 3, 4,
    6, 8, 17, 18, 19, 21,
    29, 40, 46, 47, 50, 51,
    55, 58, 59, 60, 70, 71, 73,
    74, 76, 81, 112, 113, 128,
    134, 147, 151, 156, 183,
    185, 186, 187, 192, 196,
    198, 201, 225, 227, 233
economy, 2, 8, 46, 64
England, 39, 41
Europe, 4, 49, 200
Everett Dolman, 6, 27,
    55, 56, 57
exclusion, *13*, *16*, 20,
    21, 23, 60, 114, 175,
    185, 186, 207
expanding, 3, 4, *10*, 19,
    57, 98, 134, 159, 172,
    221, 227
expansion, *1*, *2*, 3, 4, 7,
    13, *15*, *16*, 18, 20,

21, 23, 24, 25, 15, 18, 19, 28, 36, 38, 39, 52, 54, 70, 72, 73, 74, 77, 78, 80, 81, 89, 93, 102, 105, 106, 107, 109, 112, 113, 114, 117, 118, 119, 122, 123, 124, 129, 135, 136, 137, 143, 147, 152, 154, 156, 157, 159, 162, 178, 179, 180, 182, 183, 189, 197, 207, 213, 214, 226, 233
exploitation, 13, 16, 19, 21, 23, 24, 25, 30, 36, 39, 46, 47, 52, 73, 74, 78, 80, 105, 107, 111, 115, 118, 122, 123, 125, 129, 137, 143, 156, 157, 160, 165, 183, 192, 197, 207, 214, 233
exploration, 8, 2, 13, 15, 16, 18, 21, 22, 24, 36, 47, 48, 49, 55, 73, 77, 78, 88, 97, 98, 99, 100, 102, 107, 117, 135, 159, 160, 162, 168, 172, 176, 178, 187, 192, 207, 208, 213, 227, 233
exploring, 10, 22, 52, 98, 101, 134, 159
geography, 46
Geopolitik, 57
Geostationary Orbit (GEO), 69, 109

Gibraltar, 19, 23, 29, 114, 135, 183, 185, 186, 208
Go, 4, 46, 48
gravity well, 38, 40, 87, 102, 113, 121, 125, 136, 172, 204, 219
hegemon, 20, 50, 87
hegemonic, 20, 49, 71, 76
hegemony, 20, 71, 107
Helium 3, 35
Helium-3, 35, 36, 50, 126, 127, 128, 129, 213, 214
human terrain, 23
humanity, 17, 35, 38, 41, 61, 98, 117, 122, 125, 126, 127, 129, 188, 198, 207, 227, 228, 232, 233
in situ, 32, 34, 74, 76, 78, 117, 120, 149
India, 147, 148, 149, 154, 185, 194, 209
industry, 8, 8, 30, *33*, 34, 35, 39, 49, 57, 72, 73, 76, 77, 81, 101, 102, 103, 105, 106, 113, 117, 140, 144, 147, 180, 188, 214
information warfare, 79
informational, 17, 18, 71, 146, 178, 192
innovation, 2, 88, 101, 111, 138, 141, 162

International Astronautical Union, 48
iron, 9, 77, 101, 123, 124, 125, 129
island, 2, 41, 42, 54, 57, 110, 228
Japan, 4, 72, 83, 147, 148, 149, 154, 198
Jupiter, 101, 126
Lagrange points, 38, 40, 112, 168
lawfare, 79, 80
legitimacy, 46, 48, 50, 51, 52, 70
Lewis and Clark expedition, 18
littorals, 42, 43, 44, 204, 205, 221
Low Earth Orbit (LEO), 69, 110, 220, 223
Mahanian, 38, 55
Mars, 5, 13, *14*, 22, 69, 74, 89, 92, 96, 97, 98, 101, 111, 114, 116, 118, 119, 120, 121, 123, 125, 127, 129, 133, 147, 159, 162, 169, 213, 218, 233
military, 3, 8, 10, 3, 5, 3, 7, 17, 18, 21, 27, 28, 29, 31, 32, 40, 54, 55, 56, 57, 58, 59, 60, 63, 67, 71, 73, 74, 76, 77, 78, 83, 88, 90, 91, 93, 98, 106, 110, 112, 113, 128, 134, 137, 138, 141, 143, 144, 150, 153, 168, 177, 179, 180, 183, 185, 186, 187, 188, 189, 192, 193, 194, 197, 198, 199, 201, 202, 205, 206, 207, 208, 209, 211, 212, 223, 226, 227, 233
Moon, 2, 2, *15*, 18, 14, 15, 31, 36, 38, 48, 49, 50, 52, 53, 73, 74, 79, 80, 81, 82, 89, 91, 92, 93, 98, 101, 103, 105, 109, 111, 113, 116, 117, 118, 119, 120, 121, 123, 129, 130, 132, 136, 143, 144, 146, 147, 148, 149, 150, 151, 153, 154, 155, 156, 157, 159, 162, 163, 165, 166, 168, 170, 207, 211, 213, 220, 233
Moonbase, 70, 116, 118, 147, 149, 150, 152, 159, 172
Mutually Assured Destruction, 194, 195
NASA, 5, *14*, 14, 88, 89, 91, 92, 130, 135, 144, 147, 151, 154, 192, 220
NASA Artemis Program, 89, 99, 159
Navy, 4, 21, 28, 93, 137, 155, 177, 180, 193, 198, 202, 208, 218, 220, 229, 231
Near Earth Asteroid, 69, 101, 105

nuclear thermal
  propulsion, 110, 150
O'Neill cylinders, 36,
  38, 125, 168, 169, 170,
  214
Orion Program
  (Nuclear Pulse
  Propulsion), 109, 131,
  150, 154, 199, 210
Outer Space Treaty, 80,
  147, 148, 150, 200, 205
Paul Spudis, 53, 99, 118,
  130, 131, 154, 155
personnel, 3, 30, *33*,
  35, 38, 137, 152, 162,
  204
Phobos, moon of Mars,
  120, 121, 122, 123
positional advantage,
  48, 180
propellant, 9, 30, 37,
  50, 103, 111, 113, 117,
  118, 120, 121, 122, 123,
  127, 136, 150, 159, 168,
  169, 222
quantum computing,
  76, 77
Russia, 1, 2, 8, 106, 141,
  147, 149, 152, 154, 155,
  185, 204
seapower, 1, 3, 28, 39,
  41, 42, 44, 54, 72, 188,
  193, 197, 199, 227
Seapower, 41, 42, 188,
  209
silent sea, 44, 54, 111,
  114

Silent Sea, 7, 172, 234
Singapore, 19, 135
Skyhook, 119, 122, 132
Solar System, 55, 78,
  117, 120, 122, 129, 154,
  173, 193, 214, 227
South China Sea, 3, 4,
  5, 3, 4, 6, 21, 112, 181
space, 2, *3*, 4, 6, 7, 8, 9,
  10, 11, 1, *2*, 3, 5, 6, 7,
  8, 2, 5, 7, 8, 13, *15*,
  17, 21, 22, 24, 14,
  15, 19, 27, 28, 30,
  32, *33*, 34, 36, 37,
  38, 39, 40, 41, 42,
  43, 34, 46, 47, 48, 49,
  50, 51, 52, 53, 54, 55, 56,
  57, 58, 59, 60, 62, 63, 64,
  66, 67, 68, 69, 70, 71, 72,
  73, 74, 76, 77, 78, 79, 80,
  81, 82, 83, 84, 85, 86, 87,
  88, 90, 91, 93, 95, 96, 98,
  99, 101, 102, 103, 105,
  106, 109, 110, 111, 113,
  114, 116, 118, 119, 121,
  122, 123, 124, 125, 126,
  128, 129, 130, 131, 132,
  133, 134, 135, 136, 138,
  139, 140, 143, 144, 146,
  147, 148, 149, 151, 152,
  153, 154, 156, 159, 162,
  165, 166, 168, 169, 172,
  174, 175, 180, 187, 189,
  192, 193, 194, 196, 197,
  198, 199, 200, 201, 203,
  204, 205, 206, 207, 209,

210, 213, 214, 215, 216, 217, 218, 219, 220, 221, 222, 223, 224, 225, 226, 229, 231, 233

Space Based Solar Power, 69, 70, 71, 73, 76, 77, 78, 89, 96, 98, 101, 105, 109, 118, 143, 144, 146, 150, 151, 159, 162, 165, 166

Space Development Theory, 3, 10, 8, 13, *16*, 18, 61, 63, 88, 93, 98, 114, 174, 180, 207, 231, 232

Space superiority, 39

spacepower, 4, 10, 11, 7, 8, 27, 28, 39, 41, 42, 43, 44, 46, 49, 54, 56, 58, 59, 60, 62, 65, 70, 86, 87, 89, 90, 94, 108, 113, 123, 134, 138, 140, 144, 150, 152, 166, 168, 174, 187, 192, 193, 194, 195, 196, 197, 198, 199, 201, 202, 204, 205, 207, 213

Spacepower, 3, 6, 8, 27, 28, 39, 42, 46, 54, 55, 63, 65, 66, 189, 193

SpaceX, 34, 114, 138, 192, 209, 211

Spanish, 20, 183

Sputnik, 15, 55, 65, 113

strategic benefit, 19, 28, 29, 186, 189

strategy, 3, 5, 7, 9, 10, 2, 4, 5, 8, 8, 13, *15*, 20, 29, 43, 55, 57, 62, 63, 64, 68, 70, 71, 72, 75, 78, 81, 86, 87, 88, 89, 90, 91, 93, 104, 106, 107, 109, 113, 134, 135, 139, 140, 143, 144, 146, 152, 156, 159, 160, 162, 165, 168, 172, 173, 231, 233

Sun Tzu, 5, 5, 73, 140

surface to air missiles (SAMs), 203

technology, 9, 2, 30, 31, 32, *33*, 35, 39, 31, 48, 57, 59, 74, 76, 78, 81, 86, 90, 105, 106, 107, 111, 116, 126, 127, 134, 136, 139, 140, 141, 144, 146, 149, 151, 153, 156, 168, 172, 213, 225, 226

Tianhe-1, 69, 98, 159

UN, 6, 49, 211

United Kingdom, 4

US, 8, 9, 10, 11, 1, *2*, 3, 5, 6, 2, 3, 7, *14*, 17, 18, 19, 21, 23, 24, 31, *33*, 34, 43, 49, 54, 55, 56, 57, 60, 62, 64, 65, 73, 74, 75, 81, 84, 86, 87, 89, 93, 98, 101, 105, 106, 107, 110, 111, 112, 113, 121, 124, 128, 129, 134, 135, 137, 138, 139, 140, 143, 144, 146, 147, 148, 149, 151, 152, 156, 159,

162, 163, 165, 166, 168, 169, 172, 173, 177, 181, 186, 193, 194, 195, 196, 197, 198, 200, 202, 203, 204, 207, 208, 218, 220, 223, 224, 227, 229, 231, 233

US Air Force, 5, 6, 19, 20, 62, 66, 68, 83, 84, 88, 109, 110, 135, 138, 180, 194, 196, 208, 209

US Army, 7, 6, 75, 79, 84, 90, 137, 177, 202

US Army Air Corps, 6, 90

US Space Force, 6, 7, 8, 60, 62, 65, 87, 88, 90, 96, 108, 112, 135, 137, 138, 143, 165, 166, 196, 197, 198, 199, 205, 215, 233

USSR, 43, 63, 106, 110, 113, 194, 196, 200, 202

vision, 3, 8, 9, 8, 13, 52, 57, 63, 77, 90, 143, 157, 174, 194, 196, 233

Vision, 9, 8

water, 9, 37, 39, 40, 41, 42, 43, 50, 89, 91, 99, 111, 113, 116, 117, 120, 121, 122, 125, 128, 129, 130, 131, 137, 154, 174, 197, 204, 205, 221, 224

Winston Churchill, 228

World War II, 56, 72, 76, 193, 225

Table of Figures

Figure 1 - Phases of Space Development (original by author)...16
Figure 2 - Astronautics and Spacepower (original by author) ...29
Figure 3 - Artist's concept for O'Neill Cylinders (credit NASA Ames Research Center)..36
Figure 4 - Artist's concept of inside of an O'Neill Cylinder (credit NASA Ames Research Center)......................................37
Figure 5 - Chinese Space Theory (original by author)................47
Figure 6 - Triangles of Grammar and Logic with interrelations, summary of Brent Ziarnick's theory of Developing National Power in Space (original by author)..58
Figure 7 - The New Space Race, Current Situation Timeline Overview (original by author) 95
Figure 8 - Phase One of the New Space Race, First Steps (original by author)..97
Figure 9 - Phase Two of the New Space Race, Into the Deep Black Sea (original by author) 100
Figure 10 - Relative gravity well for Earth and Moon (original by author) 102
Figure 11 - Phase Three of the New Space Race, Cislunar Mediterranean (original by author)..104
Figure 12 - Phase Four of the New Space Race, The Blue Water Space Force (original by author) 108
Figure 13 - Lagrange Points relative to Earth and Moon (original illustration by author) 112
Figure 14 - Phase Five of the New Space Race, The Final Frontier (original by author) 115
Figure 15 - Artist's concept for a Skyhook (credit NASA archives) 119
Figure 16 - The New Proposed American Space Strategy showing new milestones in italics (original by author)..........................145
Figure 17 - Conceptual art of fission nuclear thermal rocket (original by author) 151
Figure 18 - The New Space Race, Reimagined, First Steps Redux (original by author) 158

Figure 19 - The New Space Race, Reimagined, Into the Deep Black Sea Redux (original by author) 161
Figure 20 - The New Space Race, Reimagined, Cislunar Mediterranean Redux (original by author) 164
Figure 21 - The New Space Race, Reimagined, The Blue Water Space Force Redux (original by author) 167
Figure 22 - The New Space Race, Reimagined, The Final Frontier Redux (original by author) 171
Figure 23 - SDT Phase 1 – Exploration, in detail (original by author) 176
Figure 24 - SDT Phase 2 – Expansion, in detail (original by author) 179
Figure 25 - SDT Phase 3 – Exploitation, in detail (original by author) 182
Figure 26 - SDT Phase 4 – Exclusion, in detail (original by author) 184
Figure 27 - Basic Diagram of Mahanian Seapower Theory (original by author) 191

Printed in Great Britain
by Amazon